THE EMERGENCE OF LIFE

The origin of life from inert chemical compounds has been the focus of much research for decades, both experimentally and philosophically. Connecting both approaches, Luisi takes the reader through the transition to life, from prebiotic chemistry to synthetic biology. This book presents a systematic course discussing the successive stages of self-organization, emergence, self-replication, autopoiesis, synthetic compartments and construction of cellular models, in order to demonstrate the spontaneous increase in complexity from inanimate matter to the first cellular life forms. A chapter is dedicated to each of these steps, using a number of synthetic and biological examples. The theory of autopoiesis leads into the idea of compartments, which is discussed with an emphasis on vesicles and other orderly aggregates. The final chapter uses liposomes and vesicles to explain the synthetic biology of cellular systems, as well as describing attempts to generate minimal cellular life within the laboratory. With challenging review questions at the end of each chapter, this book will appeal to graduate students and academics researching the origin of life and related areas such as biochemistry, molecular biology, biophysics, and natural sciences. Additional resources for this title are available online at www.cambridge.org/9780521821179.

PIER LUIGI LUISI became Professor Emeritus (Macromolecular Chemistry) at ETH-Zürich in 1982, where he also acted as Dean of the Chemistry Department; he is currently a professor of Biochemistry at the University of Rome 3. He has authored *c.* 300 papers in the fields of enzymology, molecular biology, peptide chemistry, self-organization and self-reproduction of chemical systems, and models for cells.

THE EMERGENCE OF LIFE

From Chemical Origins to Synthetic Biology

PIER LUIGI LUISI

University of Rome 3

CAMBRIDGE
UNIVERSITY PRESS

CAMBRIDGE UNIVERSITY PRESS
Cambridge, New York, Melbourne, Madrid, Cape Town, Singapore,
São Paulo, Delhi, Dubai, Tokyo, Mexico City

Cambridge University Press
The Edinburgh Building, Cambridge CB2 8RU, UK

Published in the United States of America by Cambridge University Press, New York

www.cambridge.org
Information on this title: www.cambridge.org/9780521528016

First published 2006
First paperback edition 2010

A catalogue record for this publication is available from the British Library

ISBN 978-0-521-82117-9 Hardback
ISBN 978-0-521-52801-6 Paperback

To my wife Claudia

Contents

Contents

Preface

There are already so many books on the origin of life, as listed on pages xiv–xi. Why then write another?

There are two answers to this question. The first comes from the desire to write a book for students – rather than a specialist book – in which the various phases of the transition to life would be laid out in a discursive way that illustrates the basic principles of self-organization, emergence, self-reproduction, autocatalysis, and their mutual interactions. Another important aspect of this teaching aim is to take into consideration the philosophical implications that are present, more or less consciously, in the field of the origin of life. I believe in fact that the younger generation of chemists and molecular biologists should be more cognizant of the connections between the biological and the philosophical quest, so as possibly to integrate the most basic language of epistemology, and see their science work in a broader dimension. This integration, when taken seriously, may also foster an interaction with the ethical and humanistic aspects of life. The age-old question: "what can science say about the domains of psyche, ethics, or consciousness?" is usually discarded by most scientists with a wave of the hand. This behavior is one of the main reasons why science has lost contact with the broad public – and again, it would be desirable that the younger generations take a different stand. Although this is not a central issue of this book, I hope to offer some hints on how this new approach might be defined.

While all these reasons are centered on the target of teaching, the other reason for the coming to being of this book is more subtle. It comes from the perception of a shift in the field of the origin of life, a new "Zeitgeist" (spirit of time), which makes it timely to propose a new discourse.

One aspect of the new Zeitgeist is the influence of system biology, a new operational framework where the behavior of an entire complex biological system is more important than – or as important as – the individual molecular events. Although the origin of this novel biology lies in the development of analytical tools, more than in a basic philosophical shift, the final consequence is an operational framework which is at some distance from the reductionistic approach of viewing life as a reaction based solely on nucleic acids. I believe that the exaggerated emphasis given until now to the prebiotic RNA world probably needs to be brought back into balance. And I believe the balance must be based on a more integrative view of cellular processes, even at the stage of the origin of life. Thus, I will give here proper emphasis to the autopoietic view of minimal life – which is generally not considered in other books. The latter chapters are devoted to the chemical and physical properties of compartments, vesicles in particular, and these are more technical in nature. In fact, this book suffers from that kind of heterogeneity that characterizes the field of the origin of life: on the one hand it thrives on epistemological concepts; and on the other hand it is based on experimental organic and physical chemistry. This double nature, far from being a problem, constitutes the very complexity and beauty of the field.

More generally, I will try to illustrate the different views on the origin of life and early evolution – notions like determinism and contingency will come into focus. All these scientific views are based on the postulate that life on Earth comes from inanimate matter; and a corollary of this postulate is that it might be possible to reconstitute life in the laboratory, at least in some elementary form. The ambition of understanding the prebiotic chemistry leading to the transition to life, and ultimately, to the Faustian dream of making life on the workbench, underlies the whole field – and is also the common thread of this book.

I do not know whether this dream will be fulfilled, but in closing I would like to cite Friedrich Rolle, a German philosopher and biologist, who, in 1863, writing about the hypothesis that life arose from inanimate matter, stated:

The general reasons for this assumption are really so impelling, that no doubt sooner or later it will be possible to show this in a clear and broadly scientific way, or even to repeat the process by experimentation.

This was written one and a half centuries ago and yet today we do not know whether we will ever get there. This book makes no pretence of showing the way, but as the pages unfold we will see some of the reasons why this enterprise is so difficult; and this in itself is a kind of positive knowledge.

Acknowledgments

A number of colleagues were very kind and helpful with their advice. I would like to thank Antonio Lazcano, and Albert Eschenmoser, who in different ways helped me with their frank comments; and Meir Lahav, Joseph Ribo, Jeffrey Bada, and David Deamer. Particular thanks are due to Dr. Pasquale Stano, whose help has been essential, particularly, but not only, with the bibliography; also Rachel Fajella helped with the editing of some parts of the manuscript. I am also particularly indebted to Angelo Merante for the illustrations and the formatting of the manuscript: without him, the manuscript would still be in my drawers. Last, but not least, I want to thank my students of the University of Rome 3, their positive feedback at the very early stages of the manuscript was very important.

Books on the origin of life

Bastian, H. (1872). *The Beginnings of Life*. Appleton.

Pryer, W. T. (1880). *Die Hypothesen über den Ursprung des Lebens*. Berlin.

Leduc, S. (1907). *Les Bases Physiques de la Vie*. Masson.

Osborn, H. (1918). *The Origin and Evolution of Life*. Charles Schribner and Sons.

Oparin, A. (1924). *Proishkhozhddenie Zhisni*. Moskowski Rabocii. (In Russian, translated into English as: Oparin, A., 1938. *The Origin of Life*. MacMillan).

Haldane, J. B. S. (1929), The origin of life. In *The Origin of Life*, ed. J. D. Bernal. World Publishing Co.

Bernal, J. D. (1951). *The Physical Basis of Life*. Routledge and Paul.

Oparin, A. (1953). *The Origin of Life*. Dover Publications.

Haldane, J. B. S. (1954). The origin of life. *New Biol.*, **16**, 12.

Schrödinger, E. (1956). *What is Life? And other Scientific Essays*. Cambridge University Press.

Oparin, A. (1957). *The Origin of Life on Earth*, 3rd edn. Academic Press.

Crick, F. (1966). *Of Molecules and Men*. University of Washington Press.

Bernal, J. D. (1967). *The Origin of Life*. World Publishing Co.

   (1971). *Ursprung des Lebens*. Editions Rencontre.

Fox, S. W. and Dose, K. (1972). *Molecular Evolution and the Origin of Life*. Freeman.

Orgel, L. E. (1973). *The Origins of Life*. Wiley.

Miller, S. L. and Orgel, L. E. (1974). *The Origin of Life on Earth*. Prentice Hall.

Ponnamperuma, C. (1981). *Comets and the Origin of Life*. Reidel.

Cairns-Smith, A. G. (1982). *Genetic Takeover and the Mineral Origin of Life*. Cambridge University Press.

Day, W. (1984). *Genesis on Planet Earth: the Search for Life's Beginnings*. Yale University Press.

Cairns-Smith, A. G. (1985). *Seven Clues to the Origin of Life*. Cambridge University Press.

Shapiro, R. (1986). *Origins: A Skeptic's Guide to the Creation of Life on Earth*. Summit.

Fox, S. W. (1988). *The Emergence of Life*. Basic Books.

De Duve, C. (1991). *Blueprint for a Cell: The Nature and the Origin of Life*. Portland Press.

Eigen, M. and Winkler-Oswatitisch, R. (1992). *Steps Towards Life*. Oxford University Press.

Morowitz, H. J. (1992). *Beginning of Cellular Life*. Yale University Press.

Margulis, L. and Sagan, D. (1995). *What is Life?* Weidenfeld and Nicholson.

Rizzotti, M., ed. (1996). *Defining Life*. University of Padua.

Thomas, P. J., Chyba, C. F., and McKay, C P., eds. (1997). *Comets and the Origins and Evolution of Life*. Springer Verlag.

Brack, A., ed. (1998). *The Molecular Origin of Life*. Cambridge University Press.

Dyson, F. (1999). *Origins of Life*, 2nd ed. Cambridge University Press.

Fry, I. (1999). *The Emergence of Life on Earth*. Free Association Books.

Maynard Smith, J. and Szathmáry, E. (1999). *The Origins of Life*. Oxford University Press.

Varela, F. J. (2000). *El Fenomeno de la Vida*. Dolmen Ensayo.

Willis, C. and Bada, J. (2000). *The Spark of Life*. Perseus Publications.

Zubay, G. (2000). *Origins of Life on the Earth and in the Cosmos*. Academic Press.

Schwabe, C. (2001). *The Genomic Potential Hypothesis, a Chemist's View of the Origins, Evolution and Unfolding of Life*. Landes Bioscience.

Day, W. (2002). *How Life Began: the Genesis of Life on Earth*. Foundation for New Directions.

De Duve, C. (2002). *Life Evolving, Molecules, Mind and Meaning*. Oxford University Press.

Schopf, J. W., ed. (2002). *Life's Origin, The Beginning of Biological Evolution*. California University Press.

Ganti, T. (2003). *The Principles of Life*. Oxford University Press.

Popa, R. (2004). *Between Necessity and Probability: Searching for the Definition and Origin of Life*. Springer Verlag.

Ribas de Pouplan L., ed. (2004). *The Genetic Code and the Origin of Life*. Kluwer.

Luisi, P. L. (2006). *The Emergence of Life: From Chemical Origins to Synthetic Biology*. Cambridge University Press.

1

Conceptual framework of research on the origin of life on Earth

Introduction

The main assumption held by most scientists about the origin of life on Earth is that life originated from inanimate matter through a spontaneous and gradual increase of molecular complexity.

This view was given a well-known formulation by Alexander Oparin (Oparin, 1924, 1953 and 1957), a brilliant Russian chemist who was influenced both by Darwinian theories and by dialectical materialism. A similar view coming from a quite different context was put forward by J. B. Haldane (Haldane, 1929; 1954; 1967). By definition, this transition to life via prebiotic molecular evolution excludes panspermia (the idea that life on Earth comes from space) and divine intervention. If we look at Figure 1.1 without prejudice, we realize that Oparin's proposition is extremely bold. The idea that molecules, without the help of enzymes or DNA, could spontaneously assemble into molecular structures of increasing complexity, order, and functionality, appears at first sight to go against chemical and thermodynamic common sense. This view, which modern biology generally takes for granted, appears in most college textbooks, specialized literature, and mass media. The background of Figure 1.1 is the continuity principle (Oparin, 1924; De Duve, 1991; Morowitz, 1992; Crick, 1996; Eigen and Winkler-Oswatitisch, 1992; Orgel, 1973; 1994), which sets a gradual continuity from inorganic matter to organic molecules and from these to molecular complexes, up to the onset of cellular life, with no qualitative gap between each stage. In this sense, then, the view expressed in Figure 1.1 is the modern version of a kind of spontaneous generation, although on a sluggish time scale.

In recent times, the challenges of creationists and their attacks on educational institutions in the United States led to some novel scrutiny of this view. There is nothing new in the arguments of the creationists since the writing by William Paley, the Anglican priest who became famous for having introduced one of the

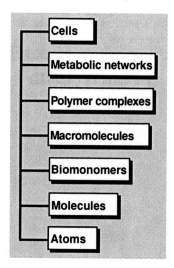

Figure 1.1 An arbitrary scale of complexity towards the emergence of life.

most famous metaphors in the philosophy of science, the image of the watchmaker (Paley, 1802):

... when we come to inspect the watch, we perceive ... that its several parts are framed and put together for a purpose, e.g. that they are so formed and adjusted as to produce motion, and that motion so regulated as to point out the hour of the day; that if the different parts had been differently shaped from what they are, or placed after any other manner or in any other order than that in which they are placed, either no motion at all would have been carried on in the machine, or none which would have answered the use that is now served by it ... the inference we think is inevitable, that the watch must have had a maker – that there must have existed, at some time and at some place or other, an artificer or artificers who formed it for the purpose which we find it actually to answer, who comprehended its construction and designed its use.

Living organisms, Paley argued, are even more complicated than watches, thus only an intelligent Designer could have created them, just as only an intelligent watchmaker can make a watch. According to Paley (1802):

That designer must have been a person. That person is GOD.

As already stated, modern science – even without reaching the extreme reductionism of Richard Dawkins and his Blind Watchmaker (Dawkins, 1990) – does not conform to this view. Paley's metaphor was already negated in his time by Hume and other contemporary philosophers. This does not mean that all scientists are necessarily atheist: the meeting point (the easy one) between science and religion is to accept the idea of a God, who created the beginning and the laws of nature, leaving them alone to do their job. We will come back to this argument a couple of times in this book.

Creationists apart, the view that life originates by itself from inanimate matter is rich with important implications for the philosophy of science and life at large. It is therefore important in our discussion to pause and consider this view, the underlying conceptual framework, as well as some of the consequences.

Let us start with the concept that is perhaps most important for lay people: it may at first sight appear that once divine intervention is eliminated from the picture, nothing remains except molecules and their interactions to arrive at life. Of course, evolution and interactions with the environment are very important factors, and they can take the fancy form of self-organization and emergence. However, all these factors appear to be based on, or caused by, molecular interactions. In other words, at first sight the acceptance of the view expressed in Figure 1.1 is tantamount to stating that life consists only of molecules and of their interactions.

Is it so? Does a rose consist only of molecules and their interactions? We can answer yes, but it is also fair to say that this would represent only a first, gross approximation. First of all, notice that the term "consists of" does not necessarily imply that life can be *explained* and *understood* in terms of molecules and their interactions. Here comes the age-old question of the discrimination between structure and properties, and whereas the structure per se can be seen as consisting of small parts, usually properties and behavior are not – or at least additional qualitative concepts are needed. In turn, this does not necessarily mean that life holds something intrinsically unexplainable or beyond the reach of science. This is an important and subtle point, and I hope to be able to offer some clarifying ideas about that in the chapter dealing with autopoiesis and cognition.

Let us consider some of the further implications of Figure 1.1. The view that cellular life can be arrived at from inanimate matter may imply in principle the possibility of reproducing it in the lab. Why not, if all we need is a bunch of molecules in a properly reactive environment? This way of thinking is the basis of the experimental work on the origin of life. In fact, the best way to demonstrate the validity of this view would be to make life in the laboratory – the age-old Faustian dream. We do not know how the process of the transition to life really occurred in nature, so how can we reproduce it in the laboratory? The answer to this question is conceptually simple, as pointed out by Eschenmoser and Kisakürek (1996):

the aim of an experimental aetiological chemistry is not primarily to delineate the pathway along which our (natural) life on earth could have originated, but to provide decisive experimental evidence, through the realization of model systems ('artificial chemical life') that life can arise as a result of the organization of the organic matter.

In other words, since we do not know, each of us is free to choose. Do as you wish so long as you show that it is possible, respecting the prebiotic conditions, to create life from inanimate matter. This is the challenge and the method is open-ended. The ambition of scientists working in the field would be simply to arrive at *minimal life*: a system containing the minimal and sufficient molecular ingredients to be called alive (this notion will be discussed in detail later on in this book). Of course this also calls into question the definition of life, a difficult issue but not an unsolvable one, as we will also see in the next chapter.

Whereas almost all researchers on the origin of life would subscribe to one form or another of Figure 1.1, with life arising from the inanimate matter, they would not agree with each other as to what is the main motor for the upward movement in the ladder of complexity. This point brings us to the next section.

Determinism and contingency in the origin of life

Is the pathway that goes from inanimate to animate matter determined by the laws of physics and chemistry? Or is it due to a unique event resulting from the contingent parameters operating in a particular time/space situation – something that in the old nomenclature would be called chance?

The dichotomy between determinism and contingency is a classic theme in the philosophy of science (see, for example, Atmanspacher and Bishop, 2002) and in this chapter it will be considered only in the restricted framework of the origin of life (see also Luisi, 2003a).

Thus, a deterministic answer assumes that the laws of physics and chemistry have causally and sequentially determined the obligatory series of events leading from inanimate matter to life – that each step is causally linked to the previous one and to the next one by the laws of nature. In principle, in a strictly deterministic situation, the state of a system at any point in time determines the future behavior of the system – with no random influences. In contrast, in a non-deterministic or stochastic system it is not generally possible to predict the future behavior exactly and instead of a linear causal pathway the sequence of steps may be determined by the set of parameters operating at each step.

Considering first the deterministic point of view, we can refer to Christian de Duve (1991); as an authorative example. In his book on the origin of life he writes:

. . . Given the suitable initial conditions, the emergence of life is highly probable and governed by the laws of chemistry and physics . . .

and later on (de Duve, 2002, p. 55):

. . . I favor the view that life was bound to arise under the physical–chemical conditions that surrounded its birth . . .

The idea of the high probability of the occurrence of life on Earth, although phrased differently and generally with less emphasis, is presented by other significant authors. For instance, H. J. Morowitz in his well-known book on the emergence of cellular life (1992, p. 12), states:

We have no reason to believe that biogenesis was not a series of chemical events subject to all of the laws governing atoms and their interactions.

He also adds, interestingly (p. 3):

Only if we assume that life began by deterministic processes on the planet are we fully able to pursue the understanding of life's origins within the constraints of normative science.

And he concludes (p. 13) with a clear plea against contingency:

We also reject the suggestions of Monod that the origin requires a series of highly improbable events . . .

This seems to lead to the idea that life on Earth was inescapable, and in fact Christian de Duve (2002), referring to a sentence by Monod to the contrary, restates this concept (p. 298):

. . . It is self-evident that the universe was pregnant with life and the biosphere with man. Otherwise, we would not be here. Or else, our presence can be explained only by a miracle . . .

Interestingly, this author, a few pages earlier (p. 289), writing about the evolution of life, has to say:

'Evolution' . . . main mechanism is by natural selection acting on accidental genetic modifications devoid of intentionality. The finding of molecular biology can leave no doubt in this respect.

This complex and apparent set of contradictions testifies to the inherent difficulties of modern scientists in having a clear-cut view of the situation.

However, as I mentioned, the idea that life on Earth can be seen as a deterministic pathway of highly probable and perhaps inevitable events is to be found frequently in the literature. In this regard, I would like to make a general point.

To say that the natural laws may have governed the prebiotic scenario and all that happened in terms of reactivity and transformations, is one thing. To say that the natural laws have constructed a series of causal steps to lead to life, is another matter; in fact, the latter assumes that the determinism is purposely guided towards the formation of life. The natural laws per se do not have a preferential direction, and actually they move without a purpose – as de Duve also mentioned above – in the direction of the most probable events. In other words, to invoke a guided determinism toward the formation of life would only make sense if the construction of life was demonstrably a preferential, highly probable natural pathway: but this is precisely what we do not know. The statement: "the origin of life must have been highly probable otherwise we would not be here" is certainly not a significant scientific statement. Rather, it is significant, only if we accept that it is based on the unconscious faith that life is unavoidable.

In fact, the same position is taken by a considerable number of the more liberal of creationists (as opposed to the biblical creationists, see Sidebox 1.1), those who accept the idea that God created the world and the natural laws, however let these laws take their own course. Thus, they can accept the science inherent in the natural origin of life, evolution, and Darwinism. Once the natural laws are given, everything develops accordingly, corresponding to a form of determinism. The problem is, that these creationists must assume that God, having created the natural laws, forcibly and purposely directed them towards the construction of life and mankind. In a way, there is an internal contradiction in this view, as one cannot invoke natural laws with corresponding determinism and then force these laws of nature into one preferential channel.

Is there an alternative to this deterministic view? One of the alternatives would be to invoke a miracle, as the one described for example by Hoyle in a famous metaphor (Hoyle, 1983): the accidental building of an airplane by a tornado whirling through a hangar full of spare parts. Rejecting this conjecture, then, de Duve (1991) claims:

The science of the origin of life has to adopt the deterministic, continuity view – otherwise it would not be possible to adopt a scientific method of inquiry,

echoing the assertions of H. J. Morowitz. This last argument – that we have to adopt the deterministic view, otherwise we are out of business – may sound naïve and tautological, but actually it is tantamount to our definition of science. Science, in its traditionalist and perhaps conservative definition, is the study and interpretation of world phenomenology in terms of the laws of physics and chemistry (with the corollary that science, also by definition, can be seen as a constant internal struggle to expand and overcome its own borders). At any rate, this definition is useful to set a clean, working benchmark between science and non-science. Science is just one part of the human enterprise, and nobody is obliged to belong to the party – but if you do it, you have to accept the more or less uncomfortable definition of science and respect the rules. At this point we should mention the "doc-creationists," those who adhere to the biblical narrative, that the world was created a few thousand years ago in seven days. One is welcome of course to have this world view, and negate all findings of science, but one cannot be a creationist and a scientist at the same time.[1] Likewise, one cannot claim to be a Christian and refuse at the same time to accept the Gospel. Either one, or the other. Sidebox 1.1, contributed by Margaret Schoeninger, shows the wide diversity of views held within the relatively small creationist movement.

Sidebox 1.1

Margaret J. Schoeninger, Professor of Anthropology
The University of California at San Diego

American creationism

In North America a strong attack is being directed toward organic evolution, especially as it relates to humans. Supported by several groups of Christians, largely outside traditionally recognized Christian religions, American Creationism is variable in its arguments although all these rely heavily on the Bible (see excellent review by E. Scott, 2004). Most emphasize biblical literalism but one subset believes Earth is young and another believes Earth is old. The former turns to the Bible for all matters including those involving the physical world. Some groups in the former subset allow for limited microevolution (within species changes) but reject all possibility of macroevolution (transformation of one species into another). For them, humans and apes have independent ancestry and Earth's geology results from a series of catastrophic occurrences like a worldwide flood. Leaders in these movements often come from technical fields like engineering (e.g., Henry Morris of the Institute for Creation Research outside San Diego, California and Walter Brown of the Center for Scientific Creation in Phoenix, Arizona).

Proponents of the second subset, which believes Earth is old (variably), include those who believe that there is a gap in time between sections of the Old Testament accounting for an old Earth, that all of geological history falls within the time before Eden, and the rest is revealed in the Bible. Others believe that the "days" described in Genesis are variable in length (>24 hours), but otherwise everything is revealed in the Bible. Progressives believe that the universe mostly developed according to natural laws, but that God intervened at strategic points along the way with regard to life on Earth. A growing, and increasingly effective group, adheres to the notion of Intelligent Design (well-funded at the Discovery Institute located in Seattle, Washington). In contrast to the other groups, individuals in this group often have post-doctoral degrees

---

[1] I believe that the main problem of the "doc-creationists" is their inability to distinguish between mythology and religion. To illustrate this I include a short personal anecdote. A few years ago I was involved in a public debate of science versus religion, in a church, with a protestant priest in Switzerland. Father S. started first, and read out to his congregation an old Sumerian legend, 600 years older than our Bible, narrating a universal flood, the birth of a child from a virgin, and other episodes very similar to those in our Bible. And then he said to his congretation: "You see, this is mythology. Let's now get to religion" – leaving me with almost no ammunition. This goes well with the statement by C. von Weizacker who said: *the Bible should be taken either seriously or literarily.*

(some in science) or other professional degrees, some from major universities. Some have faculty positions in major universities (e.g., P. Johnson, an emeritus professor of law at UC Berkeley). This view includes a supernatural, personal Creator that is proven by the presence of order and intricacy or complexity, who initiated and continues to control the process of creation toward some end or purpose. They oppose science as defined by the Arkansas balanced-treatment case in 1982, that Science is (a) guided by natural law, (b) explanatory by reference to natural law, (c) testable against the empirical world, (d) tentative in its conclusions, and (e) falsifiable.

Macroevolutionary processes are accepted in varying degrees, but the key issue is to have an involved, personal creator. In contrast to the preceding groups, one set of Creationists, including the majority of Protestant seminaries and the Catholic Church, believe in Theistic Evolution. The theory holds that there is a Creator who relies on nature's laws to bring about a purpose, that the Bible is not to be taken literally, that science is the method of choice to investigate the world, and that evolution is not seen as a contradiction to theism. In their view, science, which is materialistic in its method of investigation, is independent of the realm of ethics and morals. This latter realm is the concern of responsible social constructs, like religion.

Professor Schoeninger grew up in an academic household in extremely conservative sections of the US (South Carolina and central Florida). Including those formative years, she has lived in 11 of the 50 states. Her BA is from the Florida (southeast), M. A. from the Cincinnati, Ohio (midwest), and Ph. D. from Michigan (midwest). Her faculty positions include: Johns Hopkins Medical School (mid-Atlantic), Harvard (New England), Wisconsin (northern Midwest) and the University of California in San Diego (west coast) plus a postdoctoral position at the University of California in Los Angeles (west coast). Although her major research interest is the "evolution of human diet", perhaps this diverse background explains her fascination with American Creationism.

It is also apparent that the anti-Darwinian movement comes not so much from the present and past Pope, but rather from side-kick zealots – see, for example, the short editorial by Holden (Holden, 2005). As for myself, I would be more sympathetic towards the creationists' camp if experimental evidence were to be provided. It is not difficult to conceive what this should be: simply find *equally old* fossils of horses, dinosaurs, hominids, snails, cynobacteria, and sword fish. As long as this simple evidence is not forthcoming, it is probably safe to be scientifically very sceptical about the creationistic view (in this sense, it is almost funny that the creationists lament some small gaps in the theory of evolution). If you are interested in the creationist movement in Latin America and Mexico, in particular, see the recent article by Lazcano (Lazcano, 2005).

The interesting conjunction in de Duve's and Morowitz's view – and all the others who adhere to the deterministic view of the high probability of the origin of life – is the rejection of the miraculous scenario, and the acceptance, more or less, of the notion of the inevitability of life under the deterministic laws of physics and chemistry. I maintain that this view is similar to the (more liberal) creationistic view, although not stated expressly by those authors. I will return to this point later in this section.

The claim of the inevitability of life on Earth is criticized by some authors, for example Szathmáry calls it the "gospel of the inevitability" (Szathmáry, 2002), and Lazcano (2003) has similar views. This "inevitability" view has its counterpart in the notion that contingency is the basic creative force for shaping the molecular and evolutionistic constructs on Earth (which de Duve, 2002, dubbed "the gospel of contingency"). It should be said that de Duve accepts contingency, but in a context other than the origin of life (de Duve, 2002).

The contingency view on the origin of life and biological evolution is not new; actually is an old icon in the history of science. One may recall Jacques Monod with his *Chance and Necessity* (Monod, 1971), his colleague François Jacob with *The Possible and the Actual* (Jacob, 1982), and the books by Stephen Jay Gould, who is perhaps the most cited author on contingency in biological evolution (see for example Gould, 1989).

Contingency, in this particular context, can be defined as the simultaneous interplay of several concomitant effects to shape an event in a given space/time situation. In most of the epistemological literature this word has aptly replaced the terms "chance" or "random event" and in fact it has a different texture. In this sense, it should not be confused simply with a "highly improbable event", as mentioned above in the Morowitz citation. For example, a tile falling on your head from a roof can be seen as a chance event, but in fact it is due to the concomitance of many independent factors such as the place where you were, the speed at which you were walking, the state of the roof, the presence of wind, etc.

The same can be said for a crash in the stock market, or the stormy weather on a particular summer's day. Interestingly, each independent factor can actually be seen per se as a deterministic factor: the poor condition of the roof predictably determines some tiles sliding off and falling down. However, the fact that there are so many of these factors, each with an unknown statistical weight, renders the event as a whole unpredictable – a chance event. If the contingent conditions are changed – perhaps only one of them – the final result will be quite different. It may happen a week later, or never. It must be added that this view is not against the laws of physics and chemistry, nor is it equivalent to advancing the idea of a miracle, it is just a stochastic view of the implementation of natural laws.

However, the implications are profound. If we were to start the history of biological evolution all over again, says Stephen Jay Gould (Gould, 1989),

... run the tape again, and the first step from procaryotic to eucaryotic cell may take twelve billions years instead of two, this implies that the onset of multicellular organisms, including mankind, may have not arisen yet or may never arise. This is contingency in its clearest form. An extreme consequence of this contingency view is Monod's belief (Monod, 1971) that the human species, being a product of contingency, might just as well not have came into existence; hence the famous notion of "being alone in the universe." As a sympathizer of the importance of contingency, I wish to stress that this "being alone in the universe" should not lead one to deduce that the humanistic and ethical values are deprived of meaning, or that the sacredness of life, if you want to call it that, is impoverished. I believe in the contrary, that the values of consciousness and ethics can be arrived at from within the human construct without the need for transcendental sources.

Can one say a final word about this dichotomy contingency/determinism? it would be wise, of course, to avoid the extremes and look for a balance. The image that comes to mind is one used by Maturana and Varela (1998), when discussing the subject of biological evolution; consistently with Kimura's views on evolution, they use the metaphor of water falling from the top of the hill: the flow of water is determined by gravity, by the laws of nature. However the actual path is determined by the accidents on the ground – the trees, grooves, and the rocks encountered on the way, so that the actual downhill flow of water is a balance between the forces of determinism and contingency.

Compromises like this are always useful and make life easier. However, often they fail in the most critical situations. For example, take one fundamental question in the origin of life: is there a transcendental power behind it, or not? It would be nice to find a balance, a hybrid between Scylla and Charybdis, but, unfortunately, this is an either/or situation.

Only one start – or many?

I would now like to consider another question partly related to contingency and determinism: whether life started only once in one particular place on Earth or several times in several places. Probably most "determinists" would say that, since life has a very high probability of arising, there is no reason why it should have started only once and only in one magical place. "Only once" is a notion appealing to "contingentists": if the conditions to start life were the product of contingency – a particular set of chemicals in particular concentrations at particular temperature and pressure and pH etc. . . . – it would be almost impossible to multiply such conditions; this implies that life started only once. This argument is also connected with the question of homochirality, to be discussed later: if life started several times, each time based on contingency, then half of the time we would have one type of homochirality, and half of the time the opposite one. Does the occurrence of only one type of configuration of amino acids and sugar suggest that there was only one start? The answer of the "determinists" would be that the preferred configuration is something "ex lege" (i.e., obligatory by law), and therefore it would happen each time the same way. . . . And generally if life started many times, what then? Well, one argument says that it does not really matter, as these different initial forms of life would sooner or later enter into competition with one another, and the strongest would prevail.

There is however an extreme view of the notion of "life starting many times" that does not comply to this easy scenario. This is the view of C. Schwabe (Schwabe and Warr, 1984; Schwabe, 2001), which is highly controversial, but worth mentioning nevertheless. He starts from the hypothesis that life comes from a distribution of nucleic acids, and that this distribution was widespread all over Earth, so that there was not one, nor two or three, but multiple starts. He then goes to the extreme of saying that all species on Earth have an independent origin – a billion independent origins. He says (Schwabe, 2004):

Multiple origins means multiple species because the energy content of various combinations of nucleotides is the same, so that chemistry has no guide for the de novo synthesis of a defined specific nucleic acid that would give rise to just one species. Many new sequences will produce many origins . . .

Clearly this means a complete rejection of the fundamental Darwinian principle of common descent. Also, he rejects mutation and natural selection as the mechanisms that produced species. Is this view also contrary to the universality of biochemistry, and in particular the monophyletic origin of life, to which most biochemists today would subscribe? Probably yes; but of course if one assumes an absolute determinism, then the laws of chemistry and physics would produce the same products at each different start. This goes against the notion of "frozen accident" and the unique origin of the genetic code. So, there was never a time on Earth with only one kind of species, and the development of species was parallel rather than sequential. Of course all these ideas are substantiated by arguments and data – for these, the reader should refer to the original sources.

It should also be mentioned that Schwabe carries this view to the extreme, and he ends up arguing that life is widespread in the whole universe (Schwabe, 2002) and that the various stages of biogenesis are thus, in principle, predictable within the realms of quantum chemistry.

Having moderately pleaded for contingency rather than for determinism, I personally feel uneasy with these perspectives; however, aside from the extremes of the Schwabe scenario, the question of multiple origins for the transition to life is a valid one – yet another question, that is valid, beautiful and unanswered.

The anthropic principle, SETI, and the creationists

The notion of the inevitability of life appears to be present in science in many forms. In my opinion the anthropic principle, for example, belongs in this category. This can be expressed in different ways but the basic idea is that the universal constants, the geometric parameters, and all things of the universe are the way they are in order for life and evolution to develop (Barrow and Tipler, 1986 and 1988; Davies, 1999; Barrow, 2001; Carr, 2001). It is the *post hoc* argument that since we are so improbable, our presence must signify a purposeful universe.

The anthropic principle can be expressed in more sophisticated forms, but I believe that my simplification given above is not at all far from the target. In fact, one reads in the primary literature, for example in Paul Davies' book (Davies, 1999):

If life follows from (primordial) soup with causal dependability, the laws of nature encode a hidden subtext, a cosmic imperative, which tell them: 'Make life! And, through life, its by-products, mind, knowing, understanding . . .'

This view is held, although not always expressed as an adherence to the anthropic principle, by several authors in the field. For example Freeman Dyson (1985) writes:

As we look out in the universe and identify the many accidents of physics and astronomy that have worked together to our benefit, it almost seems as if the Universe must in some sense have known that we were coming.

We can even add a citation (Shermer, 2003) from Stephen Hawking, a self-defined atheist (in his book, the word "God" appears on almost every other page):

And why is the universe so close to the dividing line between collapsing again and expanding indefinitely? . . . If the rate of expansion one second after the Big Bang had been less by one part in 10^{10}, the universe would have collapsed after a few million years. If it had been greater by one part in 10^{10}, the universe would have been essentially empty after a few million years. In neither case would it have lasted long enough for life to develop. Thus one either has to appeal to the anthropic principle or find some physical explanation of why the universe is the way it is.

It is interesting that the anthropic principle finds more supporters among physicists than biologists, who remain in general rather skeptical about this (see for example Erwin, 2003). One general question is whether the extract by Paul Davies and his colleagues can be defined as a scientific argument – or just a claim of faith? One scientific argument that is often used by adherents of the anthropic principle is the ubiquity of biological convergence: the fact that the paths of evolution are relatively few (see for example Conway-Morris, 2003). However, it has already been shown by Gould that architectural constraints limit adaptive scope and channel evolutionary patterns, to use the wording of Erwin (2003). See also the modern extension of the anthropic principe into cosmology and the notion of "multiverse" (Livio and Rees, 2005).

There are many counter-arguments to the anthropic principle. For example, things are also the way they are in other parts of the universe, and there also slight changes in geometric distance would bring about cosmic catastrophes. Yet there is no life there. More than anything, I see in the anthropic principle a great tautology. Of course life, being life, is a granted mystery. Somebody once said "If you believe in life, then you can believe anything". This is a beautiful sentence, but we should not forget that everything that is unknown is a miracle until it is explained: lightening, the phases of the moon, the growth of the rose. In fact, one might ascribe the anthropic principle to the general category of "crypto-creationism" or more generally to the stream of the "inescapability of life" – as it implicitly contains the belief of at least an intelligent design. All this goes back to William Paley's watchmaker metaphor: as already noted, creationists or those adherent to the intelligent design have said nothing new since then. Many people see the anthropic, or egocentric, view as a position of faith – that things are the way they are in our part of the universe just to permit life. It is a view that is decidedly opposite to the principles of contingency and a view that, implicitly or not, pushes towards natural theology as an explanation for the mysteries of the universe. I repeat here my deep respect for the religious view – it is probably good to keep a little door open; but in doing this it is important not to confuse the boundaries of science and religion.

The argument of the anthropic principle – that the great laws of nature are the way they are otherwise there would be no life – is a truism at many levels. If one considers the atmosphere, there would be no life if there were more oxygen, or less oxygen; or a higher temperature, or a lower one; or less humidity, or more humidity, etc. The same is true in the molecular world. Of course if on Earth there had only been diketopiperazines and not amino acids; or if sugars did not have the size they have; or if lipids were three times shorter, then we would not have life.

This last consideration may lead to a more down-to-earth question: why has a certain type of molecular form been selected in the construction of life – and not another? I consider this type of question more scientifically sound than those of the anthropic principle, because it can be answered on the basis of thermodynamic arguments and because it permits one to perform experiments. For example, why has the five-membered ring ribose been selected out and not, for example, the six-membered piranose ring? To deal with this question in an experimental way is a constructive way of understanding the nature of life. This is the approach taken by Albert Eschenmoser and his group at the Swiss Federal Institute of Technology (ETH) of Zürich (Bolli *et al.*, 1997a, b). In one of his essays, Eschenmoser reflects on the relation between cosmic anthropic principle and the fine tuning of chemico-biological life. He considers the specific case of RNA, and writes (Eschenmoser, 2003):

... the strategy may read as follows: Conceive (through chemical reasoning) potentially natural alternatives to the structure of RNA, synthesize such alternatives by chemical methods, and compare them with RNA with respect to those chemical properties that are fundamental to its biological function.

(See also Eschenmoser, 1999). We will come back to this point in the next chapters, dealing with prebiotic chemistry.

Let us consider now another scientific movement that in my opinion seems to operate against the framework of contingency. This is the field of SETI (Search for Extra Terrestrial Intelligence), where scientists are trying to catch signals from the cosmos, believing that there is a finite probability that alien civilizations exist and are willing to communicate with us (Huang, 1959; Kuiper and Morris, 1977; Sagan, 1985; Horowitz and Sagan, 1993; Sagan, 1994; Barrow, 2001; Wilson, 2001; and perhaps you will want to reread the article by G. G. Simpson, 1973, on possible alien civilizations). The point I want to make here concerns the cultural background of this research. In fact, with the assumption of intelligent life similar to ours on other planets, the distance from contingency could not be greater. The assumption of intelligent life elsewhere is based on the unproven assumption that the same or a similar set of conditions is operative on that other (unknown) planet. Not only should one then believe in the determinism of life on our planet, but also in a kind of cosmic determinism that leads to the occurrence of life on other planets: determinism squared.[1]

Again, it is far from my intention to throw a lance at SETI. Personally, I think that this is a great vision, and that visions in science should be encouraged, particularly in an era in which mostly pragmatic and applied research projects find support. My point is rather to emphasize that this movement is also based on the belief that life is inevitable and widespread.

Conceptually close to the idea of SETI is the idea of a general panspermia, which assumes that life on Earth originated elsewhere in the universe and came to us in the form of some vaguely identified germs of life. This view has appealed already to the ancient Greeks, such as Anaxagoras (500–428 BC), right through to Hermann von Helmhotz and William Thomson Kelvin at the end of the nineteen century, to Svante Arrhenius in the beginning of the twenty century and ending with Francis Crick, Fred Hoyle and Chandra Wickramasinghe (1999; 2000) in our time; see also Parsons (1996) and, for example, Britt (2000). These different versions have different names, such as Arrhenius's radio panspermia, Crick's directed panspermia, ballistic panspermia (meteorites), or modern panspermia from comets (Hoyle and Wickramasinghe). In the more general and poetic version, the theory of panspermia sees life as a general property that permeates the cosmos and therefore

---

[1] For an interesting discussion on the relationship between SETI and ID (intelligent design), see the article by Robert Camp in the e-mail newsletter of Skeptics, www.skeptic.com, in the *Skeptic Magazine* of February 16, 2006.

does not need to have an origin (Britt, 2000; Hoyle and Wickramasinghe, 1999; Shostak, 2003). How far can one go with this idea? Wickramasinghe *et al.* recently published a paper in the well-respected medical journal *The Lancet* with the theory that SARS has a panspermia origin (see the comments by Ponce de Leon and Lazcano, 2003). There is a gray zone between science and science fiction that I personally find fascinating, but is very difficult to navigate without inhibition.

In general, on the issue of contingency versus determinism, the large majority of scientists nowadays are probably on the side of contingency. For most chemists, molecular biologists, and physicists, the notion of contingency is almost trivial. However, it is also true – as we have seen – that a significant part of the scientific population rejects the rationality of contingency and favors a determinist view of the origin of life.

How does one explain this basic dichotomy in the same generation of scientists? An easy way to describe the contradiction is to say that scientists, having pushed God out of the front door, let Him enter again through the back door. More than God per se, I believe it is the notion of the sacredness of life that has sneaked in the back entrance.

In this regard, Carl Gustav Jung's archetypes of the collective unconscious come to mind. An archetype is the part of the mental structure that is common to all mankind and that, according to Jung and his scholars (von Franz, 1988; Meier, 1992), represents the creative matrix of all conscious and unconscious functions. In their exchange of letters (Meier, 1992), the well-known physicist Wolfang Pauli and C. G. Jung discuss at length the influence of archetypical mind structures on science.

In our specific case, we would have to invoke a collective unconscious structure (the archetype of the sacredness of life?) that influences the *Weltanschauung* of scientists somehow to maintain the holy nature of life. This archetype would not appear with the same intensity in all scientists, but would be more manifest in those for example who have, or have had, a religious background. Of course, by definition of the unconscious, the beholder is not aware of his own mental behavior.

All these "crypto-creationist" movements tend to negate contingency and chance as the constructors of life and mankind, as reiterated by the following extract from Monod (Monod, 1971):

We would like to think ourselves necessary, inevitable, ordained for all eternity. All religions, all philosophies and even part of science testify to the unwearyingly heroic effort of mankind, desperately denying its own contingency . . .

I will return to the controversy between determinism and contingency in Chapter 4, taking the concrete example of the sequence of biopolymers.

Questions for the reader

1. Do you accept the view that life on Earth originated from inanimate matter without any contribution from transcendent power?
2. Do you accept the idea that biological evolution is mostly shaped by contingency? If not, what would you add to this picture?
3. Are you at peace with the idea that mankind might not have existed; and with the idea that we may be alone in the universe?
4. Do you accept the idea that a rose is made up only by molecules and nothing else?

2

Approaches to the definitions of life

Introduction

In this chapter two related questions will be considered. Firstly, the definition(s) of life; secondly, the ideas on how to implement such views on life in the laboratory. The idea of defining life is generally met with scepticism or benevolent nods by a large number of scientists. The arguments behind this negative attitude are various: such an enterprise is deemed neither useful nor easy, since everybody knows what life is, but to describe it in words is impossible. A slightly more sophisticated argument is based on the assertion that the transition from non-life to life is a continuous process, and therefore the discrimination between the living and the non-living is impracticable.

This negative attitude has several components, which have been partly analyzed (Luisi, 1998). One main problem is that the term "life" is too vague and general, and loaded with a number of historical, traditional, religious values. In particular, in the Christian tradition, the term life is generally linked to the notion of soul – and in Buddhism is linked to the notion of consciousness.

Of course, this is too much of a big picture and to define life at this level may indeed appear impossible. However, one can scientifically tackle this question by looking at life in its simplest expression, namely microbes and other unicellular organisms. This is a first, important clarification, which also eliminates (at least for most scientists) the notions of soul or consciousness from the picture. In other words, let us talk only about microbial life, and try to give a definition of cellular life.

Another difficulty in attempting to give a definition of life is that in fact the term "definition" is too ambitious, too frightening. Probably the term "description" would be more acceptable. In the language of epistemology, there is the distinction between an *intrinsic description*, meaning a context-independent description based on first principles; and an *operational description*. As Primas says in a different context (Primas, 1998):

. . . by contrast an operational description refers to empirical observations obtained by some pattern recognition methods which concentrate on those aspects we consider as relevant.

Actually, most of the "definitions" of life given in the literature comply to the above operational description. In the following pages, the term "definition" is used mostly as a way of habit, meant however in the above epistemological context.

Even so, there is another clarification to make in order to avoid further confusion on the matter. This is the following: life and its definition have been discussed on two distinct levels. On one level life is considered mainly as a *genetic population phenomenon*: one generation of *E. coli* makes the next generation of *E. coli*; a culture of green peas produces the next green peas family, and so on for all animals and plants. The alternative view of life, familiar mostly to chemists, physicists and to those in the field of artificial intelligence, is that at the level of the *single individual*. A scientist looks at a single specimen (e.g., a novel robot; or a synthetic supramolecular complex; or a single specimen of a new jellyfish; or a specimen of presumed life on a distant planet) and asks the question: is this entity actually living or not? In this case the analysis is focused on one single organism under inquiry; the historical background may be unimportant, since it may be unknown or impossible to establish. This kind of local, "here and now" view is the one that demands a criterion to discriminate between the living and the non-living on an immediate basis – without waiting for reproduction (that particular specimen may be sterile, or may take a thousand years to reproduce . . .).

However it is clear that, even when accepting these three limitations (microbial cellular life, the notion of description rather than absolute definition, and the discrimination between genetic and individual life), the question of a definition of life remains a "hairy" problem, mostly because we all have different definitions (descriptions) of life, depending on our own bias and philosophical background. I have observed – and am resigned to – the fact that it is practically impossible to bring physicists, chemists, and biologists to an agreement on what life is.

However, I still think – despite these intrinsic difficulties – that it is important to debate the question of the definition of life, both from an intellectual and a practical point of view. Everyone working in the field of the origin of life should be able to provide their own definition (or description) of life, simply because they work experimentally or theoretically on models of minimal life; and they should state and define the subject of the inquiry and the final aim of the work. This corresponds already to a kind of definition of life. Considering the number of authors in the field, this may correspond to quite a large number of different research projects, but all should adhere to the same basic constraints. What are these?

Firstly, I believe that any of the above descriptions of minimal life should permit one to discriminate between the living and the non-living. All forms of life we empirically know about should be covered by such a definition – and conversely one should not be able to find forms of life that are contradictory to such a definition. Secondly, there is the intellectual challenge to capture in an explicit formulation the quality of life: how can one express the common denominator of micro-organisms, plants, animals, mushrooms, and mammals which set them apart from the inanimate world of rocks and machines? Clearly, even if we do not arrive at an unique definition of life, the two above conditions are capable of fostering useful discussion and progress in the field.

A historical framework

After these preliminary considerations, we can look at a few definitions of life given in the literature. For a taste of them, the reader may refer to those mentioned in the monographs by Folsome (1979), Chyba and McDonald (1995), or in a book edited by the late Martino Rizzotti (Rizzotti, 1996; Popa, 2004).

In addition, a few dozen definitions of life are given in over forty pages by a corresponding number of authors, in the book edited by Palyi *et al.* (2002). They are all different from one another, some very short and some very lengthy. From each, of course, something can be learnt, but the general view is also that each one, without much comparison with previous literature and the corresponding published critical constraints to the field, pretends to have caught the truth. Is it really the case, that there are dozens of different truths on the subject? It all depends, as previously mentioned, on the meaning one wishes to attribute to the term "definition" of life: in particular if one wants to use it as an operational description to make something in the laboratory, or if one wants instead to embark on an intellectual, philosophical definition based on primary natural laws. We will see some more of these two extremes in this chapter. In the previously cited book (Palyi *et al.*, 2002) I would like to bring attention to the interesting paper by Alec Schaerer (2002), who approaches the conceptual conditions for conceiving and describing life, including the aspects of language, cognition, consciousness; and, in terms of originality of thought, also the paper by Kunio Kawamura (2002), who approaches the origin of life from the angle of "subjectivity", referring to the philosophical work by Imanishi (for me, there are strong ties with the view of autopoiesis, to come later in this book). This author provides then a view of life from the classic Japanese philosophical view, with the notion of *shutaisei* (subjectivity). In the same book there is also the Vedanta view of life (Apte, 2002) as well as that of the Russian Orthodox tradition (Arinin, 2002). There are questions about life: "Is life reducible to complexity?" (Abel, 2002); "When did life became cyclic?" (Boiteau *et al.*, 2002); "does biotic life exist?" (Valenzuela, 2002) . . . All this is just to reiterate the question about life, and the definition of life, elicits answers and input from the most differse human cultures. And from this, one should learn that in order to make sense out of these questions (what is life?; can we give a definition to life?) one should first limit the breadth of the inquiry – for example, limit oneself to defining life by looking at bacteria, and stop here before proceeding further. Go back first to the roots of simplicity. This is the approach we will follow in this book.

Now, let me go back to a few historical notes arbitrarily selected, with the aim of illustrating the historical framework (see also Luisi, 1998). Only the naturalistic view of life, excluding the creationistic or transcendental view will be considered. Let's start with the German biologist and philosopher Friedrich Rolle who noted a long time ago (1863):

The hypothesis of an original emergence of life from inanimate matter . . . can at least offer the advantage of explaining natural things by natural pathways, thus avoiding the invocation of miracles, which are actually in contradiction with the foundations of science.

This was about the same time that Darwin himself thought of a naturalistic view of the origin of life: remember his little warm pond full of salts and other good ingredients, which later on would become the famous prebiotic soup? However, Darwin didn't think too much about the origin of life. Some of the contemporary scientists who popularized his views, however, did it for him, most notably Ernst Häckel, who stressed that there is no difference in quality between the inanimate and the animate world (*Anorgane und Organismen*) and that therefore there is a natural and continuous flux from the one to the other (Häckel, 1866). This very "continuity principle" was also advocated clearly by Friedrich Rolle (Rolle, 1863; Pryer, 1880).

Proceeding with the historical discourse, let's consider a surprisingly acute definition given by F. Engels (yes, as in Engels and Karl Marx) written in 1894(!):

Life is the existence form of proteic structures, and this existence form consists essentially in the constant self-renewal of the chemical components of these structures.

This is indeed surprising, given the early date and the fact that Friedrich Engels certainly was not a great biologist, and that at this time nobody had a clear notion of what proteins really were.

We had to wait over fifty years to have a more scientific rendering of Engel's concept. Let's consider a definition written by Perret in the early 1950s, and reiterated by Bernal in 1965:

Life is a potentially self-perpetuating system of linked organic reactions, catalyzed stepwise and almost isothermally by complex and specific organic catalysts which are themselves produced by the system.

Bernal discusses this concept in more detail in his other books (Bernal, 1951; 1967; 1971). In jumping from Engels to Bernal, we should not forget a big name in between them, Alexander I. Oparin. He gave a description of life based on six properties: (1) capability of the exchange of materials with the surrounding medium; (2) capability of growth; (3) capability of population growth (multiplication); (4) capability of self-reproduction; (5) capability of movement; (6) capability of being excited. However, he also added some additional properties, such as the existence of a membrane (a cardinal principle for him); and the interdependency with the milieu (Oparin, 1961). (I would highly recommend that you read the introduction to Oparin's book, as it is still one of the best discussions on the naturalistic essence of life and the progress from non-life to life). An enumeration of properties appears to be the preferred way of getting around the problem of giving a definition in a nutshell, and modern examples of this are given by Koshland (2002) and by Oró (in Schopf, 2002).

The oldest definitions of life, such as those given by Engels, Bernal, and Rolle, referred to what we have called life at the individual level. With the advent of molecular biology and the emergence of nucleic acids, genetics also became important when considering the definition of life.

An important milestone in this regard is the so called "NASA definition of life". This was originally simply an operational perspective used by the Exobiology Program within the National Aeronautics and Space Agency – a general working definition. However those working on the origin of life often use this definition – used earlier by Horowitz and Miller (1962) – which is as follows (Joyce, 1994; my thanks to G. Joyce for this information):

Life is a self-sustained chemical system capable of undergoing Darwinian evolution.

As already mentioned, this operative definition is one of the most popular, and I must confess that I have never understood why. First of all, the notion of Darwinian evolution can only be applied to a population – therefore immediately excluding all single specimens, artifacts, chemical, and artificial life forms. Suppose that some NASA astronauts encountered single plants or single dangerous animals; they would not define them as alive, since there is no Darwinian population, and they may be eaten up in the meantime. Victims of a wrong definition of life!

More seriously, this definition is very restricted, as it is applicable only to systems that *a priori* obey the Darwinian mechanism: even the population of a cell that does not replicate, or that replicates according to a non-Darwinian scheme, would be excluded from consideration. Also, the proof of a Darwinian evolution may need thousands of years of observation – too long also for the youngest and most patient astronauts!

Of course, here is where the use of the vague term "capable of" helps, as it is probably meant to indicate that it may be just enough to look at the genetic material and see whether Darwinian-like mechanisms might, in principle, be operative – in other words, whether DNA and RNA exist, and behave as we know they should. However, then, more than a definition, we have a tautology – life must correspond to what we expect it to be.

The popularity of the NASA definition among the scientists studying the origin of life reflects this obvious prejudice, that the molecular mechanism of nucleic acids is the main and only reality for defining life. This definition is particularly dear to all adherents of the RNA-world, and actually the NASA definition appears to be created to define life at the molecular level. In fact, some of the adherents to the RNA-world would probably be satisfied by the definition of minimal life summarized (Luisi, 1998) in the following terms:

life appears as a population of RNA molecules (a quasi-species) which is able to self-replicate and to evolve in the process.

According to this view, life is made equivalent to a single molecular species, once it is capable of self-replication. It should be noted that this does not correspond to any form of life presently known on Earth and as such should rather be considered as a form of *artificial life*. The implication in the above definition is, however, that this form of life was the most likely initial source for the origin of life on Earth. Clearly, this operational definition emphasizes evolution as the main aspect of life.

This opens interesting questions, such as: are life and evolution really equivalent? Namely: if a microbial specimen does not show measurable signs of evolution, does that mean that it is not living? Suppose that you find somewhere simple forms of life that do not die and do not reproduce – would that be non-life? Also, all kinds of imputed life forms that do not have a genetic apparatus would not be contemplated by the Darwinian scheme. Would a self-reproducing cellular system with a primitive metabolism and without genes – as Oparin (1938) or Dyson (1985) have suggested – not be living? Also, a quote from Piries, 1953, comes to mind:

If we found a system doing things which satisfied our requirements for life but lacking proteins, would we deny it the title?

Considering the overlap between evolution and life, one may recall the distinction made by Szathmáry (2002) between the units of life and the units of evolution. This author emphasizes that the two domains (life and evolution) may partly overlap, but that they should be considered as two distinct realms. I have already mentioned that the NASA definition of life, based on Darwinism, is not at all meaningful when one deals with life at the individual level.

In fact, even forgetting our unfortunate astronauts, there are interesting cases where the term life may apply to a single specimen alone. The most obvious, and perhaps the most interesting for chemically oriented people, is when an artificial cellular construct is being made in the laboratory. I mention this also because it will be considered in Chapter 11, and actually most chemical laboratories are moving in this direction.

To conclude this section, let me add that there are other interesting cases that fall into the category of the single specimen: for instance, is our planet (by some considered to be self-maintaining and self-regenerating) living – as Lovelock (1979 and 1988) and Margulis and Sagan (1995) are proposing?

This section has indeed given a complex scenario of the question of the definition of life. The picture may appear confusing, and just for this reason one may ask whether or not it is not possible to go back to base, and look for one definition of life that would be simple and at the same time general enough to include individual specimens; and contain the Darwinian view as a particular case.

I believe this is possible – of course this is a bias – and in what follows this attempt will be described, based on work presented some time ago in collaboration with Varela and Lazcano (Luisi *et al.*, 1996). This is a very simple approach to the question, which in my experience is very apt for undergraduate college students.

To this aim, let us focus on the question "what is life?" from the point of view of a lay person, without knowledge of cellular or molecular biology (as certainly this question could have been answered a few centuries ago) before the advent of the microscope or of DNA. What would the answer to the question have been at that time, how to discriminate the living from the non-living? And what is the common denominator of all living things – from plants to insects to fish to people to elephants – that is not (cannot be) present in any non-living thing?

The visit of the Green Man

In order to address these questions on the definition of life first *at the individual level*, let us use a well known metaphor, utilized for example by Oparin (1953) and by Monod (1971) in another context. It goes like this: suppose than that an intelligent creature from a very distant solar system – a Green Man – comes to visit Earth in order to investigate what life is on our planet. He has has a long list of terrestrial things about which he is in doubt as to whether they are alive or not. He

Table 2.1. *The game of the two lists*

List of the living	List of the non-living
Fly	Radio
Tree	Automobile
Mule	Robot
Baby	Crystal
Mushroom	Moon
Amoeba	Computer
Coral	Paper

Question:
What discriminates the living from the non-living?

In other words:
What is the quality (or qualities) that is present in all members of the "living" list and that is not – and cannot be – present in any of the elements of the "non-living" list?

encounters an intelligent but scientifically naive farmer who very rapidly divides the items of the Green Man into two lists (Table 2.1), a list of living and one of non-living things. The Green Man is surprised by the rapidity with which the farmer operates such a discrimination, and asks him how he did it, namely he wishes to know the quality which characterizes all living things (left-hand side), and which is not present in items of the right-hand side.

When the farmer, pointing at the mule, says "movement," and "growth," the Green Man nods reservedly, as the tree or the coral in the same list do not move about, nor show any appreciable sign of growth in a reasonably long observation time; conversely, a small piece of paper moves in the wind and the moon and the tides move and grow periodically. When the farmer then gives "reaction to stimuli" as an alternative criterion, the Green Man again nods unconvinced, as the mushroom and the tree seem insensitive to a needle; and on the other hand the computer and the radio easily become ineffective upon interference with a stick.

"Living things" – adds the farmer who begins to get irritated – "are able to perform their functions by uptake of food and consequent production of energy. Energy is transformed into action." But the Green Man indicates the car, and the robot, which are able to move about by doing precisely that – converting energy into action. "Reproduction! cries the farmer. All items in his list are able to reproduce themselves!" So it is for chicken and men, but it is not so for the mule, who is unable to reproduce" – scorns the Green Man. "Furthermore, babies and very old people do not reproduce: are they not alive? Also," he says, "to reproduce it takes at least two of a kind, and I want to know about a criterion for life at the level of a single specimen".

The farmer gets more and more angry, and by doing so he arrives at a kind of enlightenment. He looks at a tree and realizes that it loses leaves in winter, but generates them again in the next spring – from the inside. Similarly it is with his beard and the hair of animals: you cut it and it grows back again – and it comes due to an activity from inside of the body!

The farmer also knows from his personal experience that when his pig is sick and fasts, its limbs and organs become somewhat smaller; however, as soon as he starts to eat again, his limbs and organs grow again. Again, this growth comes from an activity inside his own body. He concludes – and tells the Green Man – that in all elements of the left-list there are internal processes that continuously destroy and rebuild from the inside the structure itself. Living organisms are then characterized by an activity that regenerates their own components!

The farmer has finally articulated the quality which discriminates the living from the non-living! The robot, computer, radio, moon, etc. are not able to regenerate themselves from the inside. If a part of the radio breaks, the radio itself is not going to build it again. However, all items on the left-hand side of the table do have this quality: they utilize external energy to maintain their own structure, and have the ability to regenerate it from within the structure itself. This seems to be the property of life that one is looking for.

The Green Man now nods affirmatively and draws a figure on the ground (Figure 2.1). In this figure, S represents a component of the living system, which is being transformed into a product P; however, the system is able to regenerate S by transforming the entering food into S again. Actually the Green Man is rather happy about all this and, accordingly, he and the farmer make up the following "macroscopic" definition of life: *a system can be said to be living if it is able to transform external matter/energy into an internal process of self-maintenance and production of its own components.*

The farmer nods, and I doubt that he understands. However, they have arrived at a "definition" of life by using macroscopic, common-sense observations. Such a simple definition might have been derived by laymen of a couple of centuries ago; however, as it is easy to see, it is also valid for the description of cellular life. It was derived for a single specimen but is also valid for a general case, and is therefore valid both for coded life and for non-coded life. Figure 2.1 corresponds to a schematization of the behavior of a cell, and we will actually see in Chapter 8, on autopoiesis, how all of this can be organized in a more rigorous treatment of life based on a boundary and on an internal mechanism of component organization. Before proceeding, it may be useful to contrast this definition of life with alternative views.

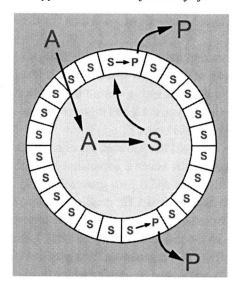

Figure 2.1 The Green Man's sketch of the living: he schematizes a system composed by only one "tissue" S; S decays into a P, but due to the internal activity of the system, the nutrient A is converted into the tissue S again. This pictorial representation corresponds to a definition of life that reads: a physical system can be said to be living if it is able to transform external matter/energy into an internal process of self-maintenance and self-generation.

Main operational approaches to the origin of life

There are of course many models and scenarios presented in the literature to explain the origin of life. Each of them represents a different way of thinking about the origin of life and about the experimental approach one should take in order to perform experiments. Rather arbitrarily, I will mention only three of these: the RNA-world, the compartmentalistic approach, and the enzyme-free metabolism approach. A similar classification is made by Eschenmoser (Bolli *et al.*, 1997a, b). A bulk of experimental and conceptual work has been collected using these three approaches. Here, I will limit myself to the description of the principles, whereas the corresponding chemistry will be considered in the next chapters. By doing so, large number of theoretical models on the origin of life will be neglected. There are more than thirty such models, and to go through all of them would certainly be instructive, but too long for the purpose of this general book. Furthermore, most of these mathematical or computer models cannot be put under experimental control, and my belief in the importance of experiments in the field of the origin of life has led to a significant bias in the literature cited.

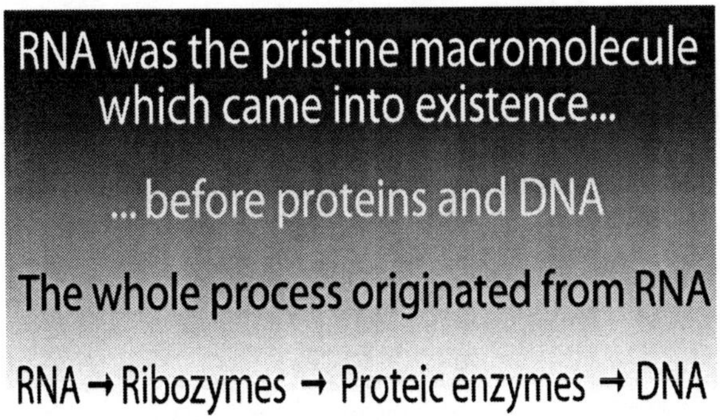

RNA was the pristine macromolecule
which came into existence...

...before proteins and DNA

The whole process originated from RNA

RNA → Ribozymes → Proteic enzymes → DNA

Scheme 2.1 The basis of the RNA world.

I. The "prebiotic" RNA world

The RNA world at large is based on the premise that RNA is the pristine macro-molecular species, from which eventually DNA and proteins are derived (Gilbert, 1986; Joyce, 1989; Orgel, 2003; Woese, 1979). This is outlined in Scheme 2.1. However, the notion of the RNA-world has acquired a twofold meaning in the literature. On the one hand there is a rich and concrete literature dealing with ribozymes, in vitro evolution of RNA, and with self-replicating RNA families – studies and data, that comprise a very important part of modern biochemistry and molecular biology. On the other hand, there is the imaginary side to the RNA world, based on the assumption that a self-replicating RNA family originated spontaneously from prebiotic chemistry – see Scheme 2.2 – and started the whole business. Such an assumption forms the basis of what can be called the "prebiotic" RNA world – a kind of intellectual by-product of the RNA world at large.

Despite an early warning by Joyce and Orgel (1986) about the molecular biologist's dream (a self-replicating RNA arising spontaneously from the prebiotic soup), this "dream" is generally found in most of the literature dealing with the origin of life on the fringes of the RNA world. It is true that there are some indications in the literature of the possible onset in vitro of ribozymes from partially random RNA libraries. Among the many examples, Joyce and co-workers (Jaeger *et al.*, 1999), Szostak and co-workers (Chapman and Szostak, 1995), Teramoto *et al.*, 2000), as well as Landweber and Pokrovskaya (1999) using a constant region and a randomised region (from 29 to 75% in the four cases), were able to detect the formation of de novo RNA with ligase activity. However, in these works RNA is not produced under plausible prebiotic conditions, rather it is obtained transcribing RNA-codifying DNA by means of sophisticated molecular enzymes such as DNA-dependent RNA polymerase. In addition, RNA molecules are not completely random due to the presence of constant regions required for the manipulation of the sample (i.e. cloning, RT-PCR and sequencining). Those are beautiful pieces of

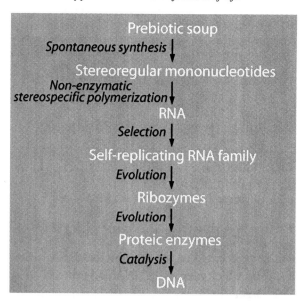

Scheme 2.2 Origin of life, in the popularized version of the "prebiotic" RNA world.

work within the RNA-world and in synthetic biology at large, but the relevance of all this for the prebiotic world is by no means granted.

This view is unfortunately also very popular in college textbooks and is often acritically accepted by most undergraduate and graduate students in the life sciences – and by inference by lay people. I say "unfortunately" because the acceptance of the spontaneous appearance of such a structurally complex and functionally sophisticated RNA molecular family is tantamount to the acceptance of a miracle – one may as well accept more traditional kinds of miracles. One should recognize that in order for real chemistry to occur many copies ($c.$ 10^{10}–10^{15}) of identical RNA molecules are needed, while one can easily calculate that the probability of quite a few identical copies of a specific macromolecular sequence capable of self-replication arising spontaneously from the mixture of monomers is essentially zero. Aside from mathematical considerations, the chemical evidence is that there is until now no ascertained prebiotic synthesis for mononucleotides. Notably, this is still the case, despite the tremendous effort of many brilliant chemists over more than 30 years of investigations. Even if the prebiotic formation of monomers were known, we would not know how to perform an enzyme-free long-chain polymerization, and even less how to make an enzyme-free 3′–5′ stereospecific polymerization. Then there is the problem of making many identical copies of the same sequence. In Chapter 7, which looks at self-replication, additional reasons are discussed as to why the idea of the spontaneous birth of a self-reproducing RNA is at the present stage not tenable.

There is of course still the possibility that some brilliant chemist will soon discover a prebiotic scenario for making RNA sequences – in a way we all hope that this will be the case, it would indeed be a good day for those studying the origin of life. However, for the time being, the "prebiotic" RNA world is grounded on the above-mentioned dream, and not on solid science.

Why then is there this popularity of the prebiotic RNA world? There are three reasons that come to mind. One is the already mentioned great success of the RNA world at large, which, by inference, gives confidence in the power of RNA. Another reason is that from self-replicating and mutating ribozymes, one can conceive in paper a route to DNA and proteins – and then one has the whole story. A third reason is the lack of a good competitive model – namely the fact that there is no alternative mechanism that is supported experimentally.

All this is no reason to write college textbooks in which a random polymerization of nucleotides magically produces self-replicating ribozymes. There are more and more critics nowadays of this "prebiotic" RNA world, but there is a general consensus that RNA was a key molecule for determining – if not the origin of life – early evolution. We will come back to this point later on in this book. I would like now to conclude this section on RNA with something positive, namely two important lessons. One is about the importance of self-replication as a basic mechanism for the beginning of the mechanisms of life. The second, more sophisticated, point, is the recognition that the search for macromolecules that contain both genetic information and catalytic power would greatly simplify any scenario concerning the origin of life.

II. The compartmentalistic approach

The main conceptual tool of the prebiotic RNA world is self-replication. The main conceptual tool of the compartmentalistic approach – if such a word does not exist, we need to create it . . . – is so that everything starts within a closed spherical boundary. The argument in favour of this view is very basic: all life on Earth is cellular, namely based on closed compartments. If we take all cells of this world – the argument sets forth – and squeeze their content into the vast ocean, we will have all the RNA and DNA in the world – and no life. Also inside the cells there are other kinds of mutually interacting compartments, and in fact the main functions of life can be seen as an interaction between the inner world and the external medium; such functions are guaranteed by the flux of material and information through the boundary of the compartments. If such an asset is so important for life, the compartmentalistic approach then suggests that in order to consider the origin of life we have to start from a primitive semi-permeable closed boundary. This scenario is reinforced by the fact that cell-like compartments – e. g., vesicles – form spontaneously with molecules of prebiotic origin, as illustrated in Figure 2.2.

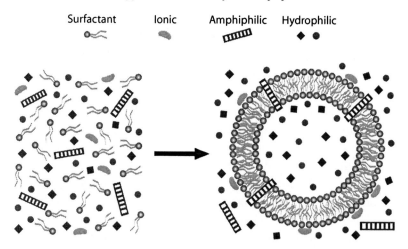

Figure 2.2 The spontaneous self-aggregation of membranogenic surfactants into a vesicle, with an interior water pool that can host water-soluble molecules. If this self-aggregation takes place also in the presence of hydrophobic molecules, and/or ionic molecules, these can organize themselves into the bilayer or on the surface of the vesicle. A realistic scenario of the emergence of life can be based on a gradual transition from random mixtures of simple organic molecules to spatially ordered assemblies, displaying primitive forms of cellular compartmentation, self-reproduction, and catalysis.

This kind of approach, although with different wording and emphasis, has been in the literature for some time. Actually the first books on the origin of life emphasize the importance of a cell-like compartment as the prime act of the evolution that eventually leads to the self-reproducing cells; Oparin (1953), Morowitz (1992) and Dyson (1985) all discuss this concept. Several other authors have worked on these ideas, most notably David Deamer (for example Deamer, 1985 and 1998; Pohorille and Deamer, 2002); my own group (Oberholzer *et al.*, 1995, 1999; Luisi and Oberholzer, 2001; Luisi, 2002a); Lancet and his group (Segre and Lancet, 2000; Segre *et al.*, 2001); as well as the groups of Ourisson and Nakatani (Nomura *et al.*, 2002; Takakura *et al.*, 2003); of Noireax and Libchaber, 2004; and of Yomo (Ishikawa *et al.*, 2004). From all this work, as discussed later on in detail in this book, it appears that it is possible to insert enzymes, nucleic acids and other biochemicals inside these compartments, so as to have in principle the beginning of a possible metabolic pathway. We will also see in detail that such compartments (micelles and vesicles) are capable of an autocatalytic self-reproduction mechanism.

The basic working idea of compartmentalism is that these primitive shells have encapsulated some simple peptide catalysts together with other molecules, and that a primitive protocell metabolism may have started in this way. However, the question of how the primitive metabolism really started is still unanswered – and in particular how that particular metabolism could have started, that led the way to

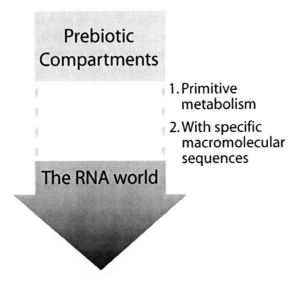

Figure 2.3 The relation between the compartmentalistic approach and the prebiotic RNA world; and the missing link (items **1** and **2**).

RNA. Thus, the microcompartment approach seems capable of giving the spontaneous foundations of the elementary basic protocells, but presently cannot proceed further; the prebiotic RNA approach appears to start from the roof (from already formed RNA molecules) and is unable to explain how this roof came about – as illustrated in Figure 2.3.

It seems that there is a missing link between the scenarios of the prebiotic RNA-world and the compartmentalistic approach: and the missing link is how to make RNA by a prebiotic sequence or network of internalized reactions.

III. The "prebiotic metabolism" approach

In a way, the missing link appears to be the aim of a few distinct research groups, who are inquiring into the possibility of prebiotic metabolic pathways prior to enzymes. There is a rather complex kaleidoscope of authors and views, and it is useful to distinguish between the following main trains of thought.

The universal metabolism

In a rather original approach, H. J. Morowitz and co-workers (Morowitz *et al.*, 1991; 1995; and 2000) examine the chemistry of a model system of C, H, and O that starts with carbon dioxide and reductants and uses redox couples as the energy source. To investigate the reaction networks that might emerge, they start with the large database of organic molecules, *Beilstein on-line* (www.beilstein.de); and on the basis of certain assumptions, from the 3.5 million entries in Beilstein they notice the emergence of only 153 molecules that contain all 11 members of the reductive citric-acid cycle. According to the authors, these calculations suggest that the metabolism corresponds to a universal pathway chart and is therefore central to the origin of life. Enzymes would have come only later and accelerate the cycles, eventually taking over; the bottom line is a metabolism prior to the origin of catalytic macromolecules.

It must be said that this view is not universally accepted; for example Orgel has argued forcibly against the Beilstein approach and generally against the metabolic cycles (Orgel, 2000). His point is that theories involving the organization of complex, small-molecule metabolic cycles, such as the reductive citric-acid cycle on mineral surfaces, have to make unreasonable assumptions about the catalytic properties of minerals and the ability of minerals to organize sequences of disparate reactions. Another conclusion is that data in the Beilstein *Handbook of Organic Chemistry* (see, for example, www.indiana.edu/~cheminfo/cciim33.html), which are claimed to support the hypothesis that the reductive citric-acid cycle originated as a self-organized cycle, can more plausibly be interpreted in a different way.

Metabolism on clay and mineral surfaces

When considering clay and minerals one should necessarily mention the pioneer work by Cairns-Smith (Cairns-Smith and Walker, 1974; Cairns-Smith, 1977, 1982; Cairns-Smith *et al.*, 1978, 1992; 1982; 1990, see also Bujdak *et al.*, 1994). This author, developing an earlier idea suggested by Bernal, showed that clay mineral surfaces could adsorb organics, building up sufficient concentrations of these as well as acting as templates for polymerization in an aqueous environment. The "primeval soup" would thus become a concentration of organic compounds not in solution but absorbed by mineral surfaces. The original intuition was probably that the replicator-catalyzing agents were actually crystals to be found everywhere in the clay that lay around primitive Earth. Crystals are structurally much simpler than any biologically relevant organic molecule, and they appear to grow and reproduce by breaking into smaller seeds that can grow further. In this way, they can be said to carry information for building themselves; furthermore, crystals can incorporate impurities while growing – a kind of primitive mutation and evolution. The next step is to propose that these very primitive "organisms" started incorporating peptides found in the environment, and the road was then open to a gradual increase of complexity of the incorporated biological world – and from there a genetic takeover.

All this sounds more like a metaphor and this is probably why the idea was so catchy – it draws on visual evidence of simple and well-known mechanisms. However, from a metaphor to a chemical reality there is a wide gap – how proteins and nucleic acids are built and/or the lipid metabolic compartment – and in fact from the experimental point of view there is very little that substantiates this kind of clay scenario. This view of Cairns-Smith has taken a good niche in the literature of the origin of life and is being constantly refreshed and reconsidered (Cairns-Smith, 1978; 1982). Of course the chemistry on clay and mineral surfaces as primitive forms of matrices for primitive metabolism has eventually to evolve into the world of membranes and lipids.

Cairns-Smith's original approach may sound different from Morowitz's approach mentioned earlier; but in fact it also postulates a kind of metabolism without and prior to enzymes.

The beauty of pyrite

Along similar lines to Morowitz is the work of G. Wächtershäuser, (Huber and Wächtershäuser, 1997; Wächtershäuser, 1990a, b; 1992; 1997; 2000). The basic idea is the proposal of an autotrophic emergence of life, based around a reductive Krebs cycle (enzyme-free of course) "running in reverse" (Lazcano, 2004), with the synthesis and polymerization of organic compounds taking place on the surface of pyrite in extremely reduced volcanic settings resembling those of deep-sea hydrothermal vents. The organic compounds formed from the reduction of CO_2 evolved into an autocatalytic, two-dimensional, pyrite-driven chemolithotrophic metabolic system, which lacked a genetic system.

Independently, there was the discovery in 1979 of the richness of organic compounds in hydrothermal hot vents (see for example Holm *et al.*, 1992, and Chapter 3). The idea was fully developed by Wächtershäuser (1988) and Cairns-Smith *et al.* (1992), and (of course) became another "world." Life then began with the reduction of CO_2 and N_2 coupled with the reducing power of pyrite formation – and so was born the "iron–sulfur-world" hypothesis. Thus, the work of Wächtershaüser also represents a link between the field of surface catalysis and the field of hydrothermal vents.

The theory is based on the autotrophic metabolism of low-molecular-weight constituents in an environment of iron sulfide and hot vents. Figure 2.4 gives an illustration of one reaction pathway. It is worthwhile to consider that the metabolism is a surface metabolism, namely with a two-dimensional order, based on negatively charged constituents on a positively charged mineral surface. Actually Wächtershäuser sees this as an interesting part of a broader philosophical view (Huber and Wächtershäuser, 1997).

The author also makes a strong point to indicate the difference between this approach and the "prebiotic soup" approach, and argues (Wächtershäuser, 2000):

It is occasionally suggested that experiments within the iron–sulfur world theory demonstrate merely yet another source of organics for the prebiotic broth. This is a misconception. The new finding drives this point home. Pyruvate is too unstable to ever be considered

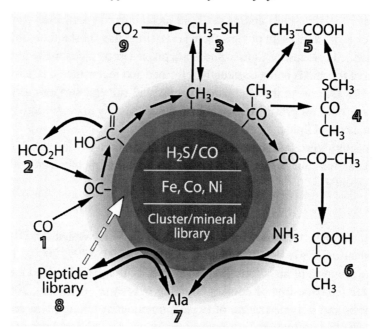

Figure 2.4 Reactions in the iron–sulfur world. The dotted arrow represents ligand feedback. For more details see Wächtershäuser, 2000, from which this figure is taken (with some modifications) with kind permission of the author. For the precise meaning of the numbers in the figure, see the original paper.

as a slowly accumulating component in a prebiotic broth. The prebiotic broth theory and the iron–sulfur world theory are incompatible. The prebiotic broth experiments are parallel experiments that are producing a greater and greater medley of potential broth ingredients. Therefore, the maxim of the prebiotic broth theory is "order out of chaos." In contrast, the iron–sulfur world experiments are serial, aimed at long reaction cascades and catalytic feedback (metabolism) from the start. The maxim of the iron–sulfur world theory should therefore be "order out of order out of order."

Several authoritative authors in the field have not spared their criticism on these views, e.g., Stanley Miller, Christian de Duve (de Duve and Miller, 1991), and Leslie Orgel. One argument by Leslie Orgel (2003) is that this "pyrite metabolism" might reflect not a pyrite-dependent primordial metabolism but the unique, deterministic way in which a given chemical processes can take place. In addition, Orgel (2000) had already underlined that, in general, theories advocating the emergence of complex, self-organized biochemical cycles in the absence of genetic material are hindered, not only by the lack of empirical evidence, but also by a number of unreasonable assumptions about the properties of minerals and other catalysts required spontaneously to organize such sets of chemical reactions.

Aside from this, and aside from the criticisms by Orgel, de Duve and Miller, the ideas have induced considerable interest and popularity – see for example a recent book on the origin of life (Day, 2002). Furthermore, let me conclude this section on pyrite with a comment by Lazcano (Miller and Lazcano, 2002) in response to Day (2002):

Compared with the surprising variety of biochemical compounds that can be readily synthesized in Miller-type one-pot simulation experiments, the suite of organics produced under the conditions proposed by Wächtershäuser is quite limited. However, the impressive demonstration that the FeS/H$_2$S combination can reduce nitrogen to ammonia shows that considerable attention should be given to the reducing power of pyrite formation. Primordial life may have not been autotrophic, but should we hesitate to accept the idea that the primitive soup was formed from both extraterrestrial sources and endogenous synthesis in which pyrite production played a role? After all, a spicy, thick bouillon is always tastier than a bland, diluted broth.

Other metabolic approaches

To the list concerned with enzyme-free metabolism should be added the work by August Commeyras and his coworkers in Montpellier (Taillades *et al.*, 1999; Plasson *et al.*, 2002). This school has described a molecular engine mechanism which could have taken place on primitive beaches in the Hadean age, a primary pump which relies on a reaction cycle made up of several successive steps, fed by amino acids, and fuelled by NO$_x$ species. For the proposed mechanism to work it was assumed that there was a buffered ocean, emerged land and a nitrosating atmosphere. The French authors argue that this primary pump might have been the prebiotic mechanism that gave rise to oligopetide sequences, which in turn started the origin of life.

Probably the "thioester world" of de Duve also belongs to this category of work relying on prebiotic metabolism without the help of enzymes (de Duve, 1991). The ideas of Shapiro on the origin of life (Shapiro, 1986) have also been influential. Of course this list is not exhaustive, theories on the origin of life abound in the literature, and here I wanted to present a quite general "spaccato" without giving a detailed review.

Perhaps one last word is needed about more theoretical work – space does not allow for a more detailed discussion. An influential author in this respect is Stuart Kauffman; in his well known article (Kauffman, 1986) he connects the origin of life to the spontaneous rise of a catalytic set of peptides, an emergence that is seen as an inevitable collective property of any sufficiently complex set of polypeptides. The main conclusion of this analysis is the suggestion of the emergence of self-replicating systems as a self-organizing collective property of critically complex protein systems in prebiotic evolution. Similar principles may apply to the emergence of a primitively connected metabolism. The question, whether and to what respect this view connects with the experimental findings cited in this chapter, is still open.

This argument about experimental work leads us to the next chapter, which gives a short analysis of prebiotic chemistry – and beyond.

Concluding remarks

This chapter began with the question of the definition of life, and two contradictory results have been obtained. On the one hand, it was argued that probably no general definition can be found, as it is practically impossible to bring different scientists and backgrounds – physicists and biologists, chemists and engineers, geologists and bio astronomers, computer experts and paleontologists – to a common conceptual denominator. On the other hand, it has been stressed that it is important to give such a definition, particularly to clarify the aims of one's work and comparison with that of others. For me, for example, the best operational definition of life is the one obtained by the theory of autopoiesis – as expressed in a primitive form by the Green Man, and more extensively in Chapter 8, on autopoiesis. As a consequence of this view, I see the construction of the minimal cell (see Chapters 10 and 11) as the implementation of such a cellular definition of life.

With regard to the various hypotheses and approaches to the origin of life, the short "spaccato" in the second half of this chapter is enough to show the large variety of views and philosophies on the matter. We have seen that "somebody likes it hot", but there is also some skepticism about the hot origin of the beginnings of life – actually "somebody does not like it hot" (Miller and Lazcano, 1995; Beda and Lazcano, 2002). Also, a well known molecular biologist, Christoff Biebricher, recently reported on the importance of ice for the origin of life (Trinks *et al.*, 2003).

The fact that we have so many different views indicates the obvious: that we do not have a convincing view that is ascertained and accepted by the majority of researchers. Partly, these views are seen in the literature as opposing each other – I have experienced fierce fights between those in the RNA world and those advocating autopoiesis/compartimentalism. As seen from Wächtershaüsers polemic against those who believe in the "prebiotic soup" (are there any around?) it is obvious that these views are complementary. Figure 2.3. is a representation of this and of the need of working in concert to build the entire house – as the roof alone (the RNA world) or the foundation alone (the compartmentalists) or the metabolism alone do not build anything solid.

In addition, as we will learn in the next two chapters, there is still something important missing. In fact, these views refer mostly to the world of low-molecular-weight compounds, namely the bricks for making the house. However, you can have all the low molecular weight compounds in the world, made by hydrothermal vents or by pyrite or by clay or by primitive metabolisms – and you do not make life with that. To make the house, you need at least the macromolecular specific sequences of enzymes and nucleic acids. This leads us nicely into the next two chapters.

Questions for the reader

1. Do you believe in the utility attempting to give a definition of life? If not, how would you answer the following questions?
2. Is an apple – hanging on a tree – living? When it falls to the ground – is it still living?
3. What is the difference between a living horse and a one which has just died? They have the same amount of RNA, DNA, and all nucleic acid reactions are working for a while. Why is the dead horse dead?
4. Take the nucleus out of an oocycte, as in the cloning experiments. Is the nucleus living? And is the cell, without a nucleus, alive?

3

Selection in prebiotic chemistry: why this . . . and not that?

> . . . l'uovo cadde dal ciel
> e come a Dio piacque
> l'uovo si ruppe
> e la gallina nacque.
> (Tuscan folkloric poetry)

Introduction

Having considered in the previous chapters the "software" of the origin of life, we are now ready to look at the "hardware" – some chemistry facts. Let us start from the beginning.

A way to connect software and hardware is summarized in Figure 3.1, which shows the most common beliefs and assumptions of researchers in the field. Particularly important is the assumption that some form of minimal life can be made in the laboratory, once the right conditions of prebiotic chemistry are found.

Life on Earth may have started between 3.5 and 3.9 billion years ago, as shown in Figure 3.2. The oldest microfossils were described by J. W. Schopf from the Apex Chert at Marble Bar, Western Australia (Schopf, 1992, 1993, 1998). These are dated at 3.465 Ga. Microfossils from Swaziland (South Africa) are of similar age. The North American Gunflint Chert (2 Ga) and the Belcher Group microfossils from Canada are the first occurrence of the Precambrian.[1]

Schopf's data have been criticized (Brasier *et al.*, 2002) and the controversy is still open. However, most scientists accept the idea that micro-organisms already existed about 3.4–3.5 billion years ago.

If these fossils were cells with an already fully fledged genome, then the origin of life must be placed earlier – and by so doing we come very close in geological time

---

[1] For more information about the microfossils discovered and published up to 2003, see: http://www. unimuenster.de/GeoPalaeontologie/Palaeo/Palbot/seite1.html

1. Life originated from inanimate matter as a spontaneous and continous increase of molecular complexity. Chemical continuity principle – no transcendental principle.

2. The chemical process(es) of transition to life can be reproduced in the laboratory with the presently available chemical techniques and chemicals.

3. This can be implemented in a reasonable (hours or maximum days) experimental time span – once you know the right combination of prebiotic compounds and the conditions.

4. Since there is no documentation on how things really happened, there is no obligatory research pathway.

Figure 3.1 Some main assumptions of present-day research on the origin of life.

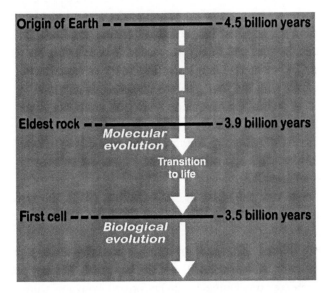

Figure 3.2 A simplified time flow of the origin of life. The solar system is supposed to have formed 4.5 billion years ago, around 9 million years after the big bang. Apparently the Earth was cold enough to host the first mild organic reactions around 3.9 billion years ago, and the first fossils are 3.45 billion years old. (Adapted from Schopf, 1993.)

to the formation of the first cold rocks. This is often taken by some as an indication that life on Earth arose *"as soon as it could"*, which in turn is taken by some to signify that the origin of life was an easy process based on robust reactions.

The idea of describing the chemical reactions that are germane to the origin of life has a long history – particularly if one starts from Wöhler's experiments in 1828 on the synthesis of urea, as recently proposed by Bada and Lazcano (2003). Wöhler's reaction was a fortunate accident, in the sense that the German chemist did not have in mind any ambitious Faustian dream. In what follows some basic notions of prebiotic chemistry are reviewed, although not exhaustively. For more extensive reviews, see Oró (Oró, 1960; 1994; 2002; Oró and Kimball, 1961), Shapiro (1988), Brack (1998), Miller (in Brack, 1998), Commeyras and co-workers (Taillades *et al.*, 1999; Plasson *et al.*, 2002; Commeyras *et al.*, 2005), Miller and Lazcano (2002), and Schopf (2002). Sidebox 3.1 by Antonio Lazcano gives some additional insights into the matter.

From Oparin to Miller – and beyond

The first scientist who consciously did chemistry in pursuit of the origin of life was the graduate student Stanley Miller, after he had read and digested the English translation of Oparin's book. How the young Miller arrived at celebrity is now part of the narrative of the origin of life, see also Lazcano and Bada, 2003. Fascinated by Oparin's idea, he tried and finally succeeded in convincing his already famous mentor Harold C. Urey to give him some time to try an experiment.

He filled a flask with the four gaseous components assumed by Oparin to be the constituents of prebiotic atmosphere: hydrogen, ammonia, methane, and water vapour – a reductive atmosphere. By passing electrical discharges through this flask to simulate primordial lightening, Stanley Miller was able to witness the formation of several α-amino acids and other relatively complex substances of biological importance (see Figure 3.3).

The experiment was published in 1953 (Miller, 1953), the same year as the discovery of the double helix by Watson and Crick, a memorable year indeed for biochemistry.

There is only limited agreement on whether primitive atmospheric conditions were really reductive. However this is not the key point. The key point is the fact that relatively complex biochemicals can be formed from a mixture of very simple gaseous components in a chemical pathway that can indeed be regarded as prebiotic.

Following Miller's experiment, many chemists successfully synthesized other compounds of biochemical relevance under prebiotic conditions – thereby demonstrating convincingly that several molecular bricks of life might have been present on prebiotic Earth.

The bases of nucleic acids can also be considered prebiotic compounds. A possible prebiotic route to adenine has been described (Oró, 1960; Oró and Kimball, 1961; 1962; see also Shapiro, 1995), as shown in Figure 3.4. For details see also Miller's review (Miller, 1998). Guanine, and the additional purines such as hypoxanthine, xanthine, and diaminopurine could also have been synthesized by variations of the above synthesis (Sanchez *et al.*, 1968).

The prebiotic synthesis of pyrimidines is based on cyanoacetylene, which is obtained in good yields by sparking mixtures of methane and nitrogen; and by the

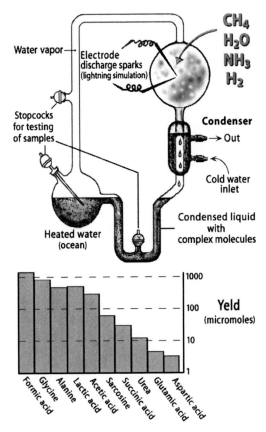

Figure 3.3 The famous Stanley Miller experiment: a strongly reducing atmosphere consisting of the four gases, and electric discharges as energy source.

reaction between cyanoacetylene and cyanate (Sanchez *et al.*, 1966; Ferris *et al.*, 1968 – see Miller, 1998). This is shown in the Figure 3.5, taken from Miller (1998), which also shows that cytosine can be converted into uracil.

It should be kept in mind that these prebiotic syntheses concern only the bases, and not the mononucleotides of the nucleic acids. The mononucleotide consists of three moieties attached to each other – the base, the phosphate, and the sugar – and we do not yet know how this prebiotic synthesis may have taken place.

With regard to sugars, a well-known reaction is the formose reaction (see Figure 3.6), an autocatalytic process that, starting from formaldehyde, proceeds through glycolaldehyde, glyceraldehyde, to reach four-carbon and five-carbon sugars. This reaction, however, gives a wide variety of sugars, both straight-chain and branched. Ribose occurs in the mixture, but is not particularly abundant (Decker *et al.*, 1982) – and furthermore this compound is chemically rather instable (Miller,

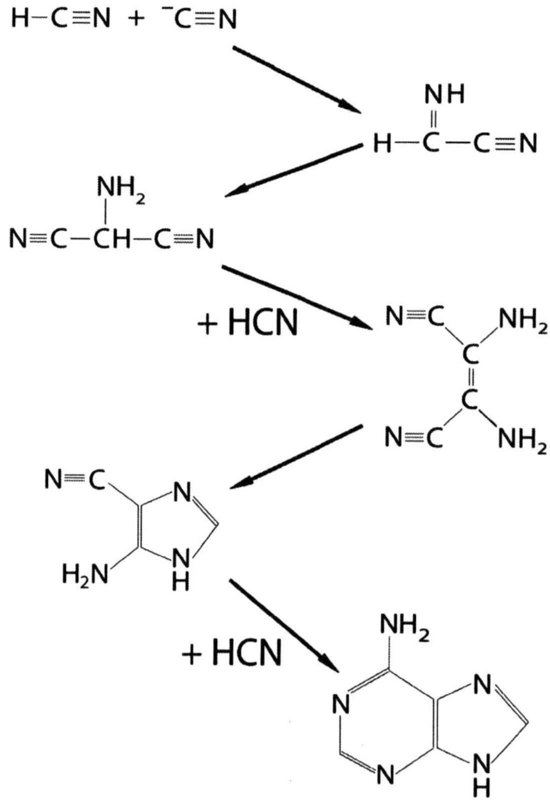

$$H-C\equiv N \ + \ ^-C\equiv N$$

Figure 3.4 The prebiotic synthesis of adenine. (From Oró, 1960.)

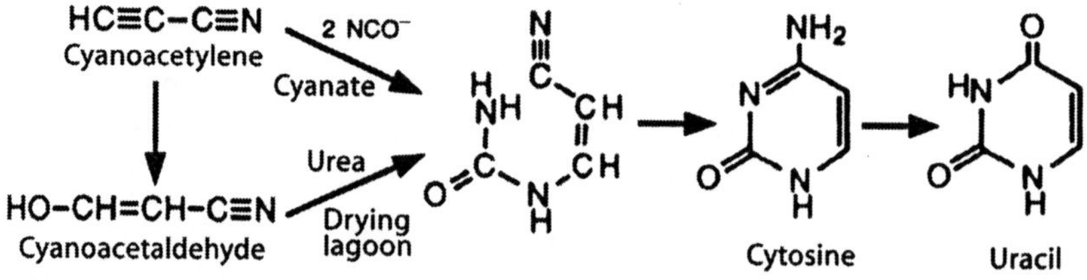

Figure 3.5 The prebiotic synthesis of pyrimidines of nucleic acids. (From Miller, 1998.)

1998). It is also worth mentioning the formation of ribose-2,4-diphosphate obtained by Eschenmoser's group (Müller *et al.*, 1990).

In his review, Miller (1998) recalls other compounds that have been synthesized under primitive Earth conditions, such as di- and tri-carboxylic acids, C_2–C_{10} fatty acids, porphin, imidazoles, and he also mentions products that have not been

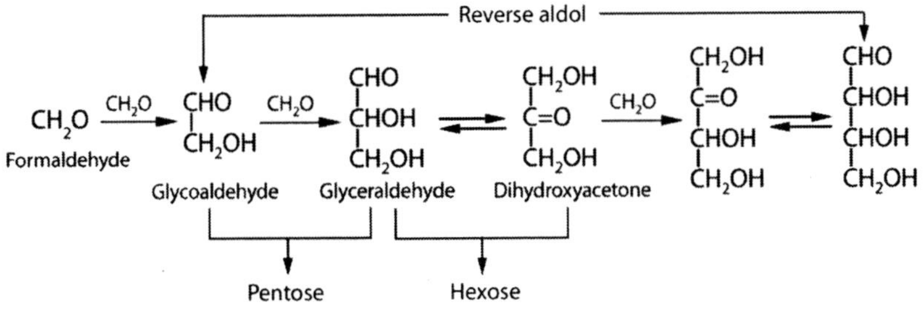

Figure 3.6 The formose reaction. (Adapted from Miller, 1998.)

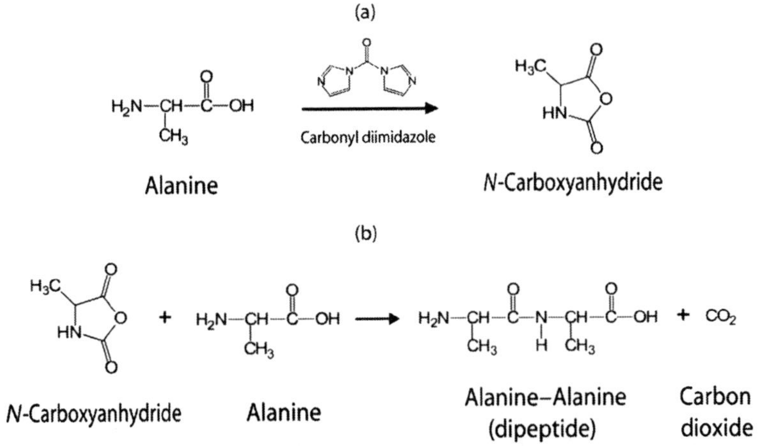

Figure 3.7 (a) Activation of amino acids by *N*-carboxyanhydride and (b) their condensation. (Adapted from Ferris, 2002.)

synthesized prebiotically, such as certain amino acids (arginine, lysine, histidine) – as well as thiamine, riboflavin, etc. For amino acids, the observation by Commeyras *et al.*, 2005 on the prebiotic nature of *N*-carboxy anhydride (NCA) is particularly important. In fact, NCA-activated amino acids are capable of condensation, as shown in Figure 3.7. For the chemistry of the NCA (or Leuchs anhydrides) see also Kricheldorf, 1990; Kricheldorf and Hull, 1979, and Karnup *et al.*, 1996. In the next chapter on macromolecules we will come back to this reaction.

To this short review of prebiotic reactions the work on prebiotic membrane-forming compounds should be added. This will be considered later on in the chapter in the section on surfactant self-organization.

Now let's return to Wächtershäuser's work, mentioned in Chapter 2, and consider it from a prebiotic-chemistry perspective, given the possible source of reductive power in a prebiotic scenario. Wächtershaüser and coworkers indicated that the reaction of FeS with H_2S, Equation (3.1),

$$FeS + H_2S \Rightarrow FeS_2 + 2e^- + 2H^+ \tag{3.1}$$

has a reducing power to drive the primordial metabolism. In fact, it has been shown that a number of non-spontaneous reactions can be coupled to the previous redox process, including the reaction of CH_3SH, from which thioesters such as CH_3COSH can be synthesized (Heinen and Lauwers, 1997; Huber and Wächtershauser, 1997). In turn, from the latter compounds other interesting and more complex molecules can be obtained, for example fixation of CO_2 (Nakajima *et al.*, 1975), Equation (3.2).

$$CH_3COSH + CO_2 + FeS \Rightarrow CH_3COCOOH + FeS_2 \tag{3.2}$$

Many researchers are excited about this pyrite scenario as a source of important metabolites. For the implications of this work as a whole, the reader is referred to the previous chapter.

The Stanley Miller experiments produce amino acids. What about peptides? In fact, there have been some investigations of the possible prebiotic origin of simple dipeptides. Shen's group reports on the prebiotic synthesis of histidyl-histidine; and this is interesting in view of the catalytic properties of this molecule (Shen *et al.*, 1990a; b). Studies on the prebiotic synthesis of histidyl-histidine were carried also out by Oró and coworkers (1990). Glycine and diglycine were investigated as possible catalysts in the prebiotic evolution of peptides (Plankensteiner *et al.*, 2002). Regarding the catalytic properties of simple dipeptides, one should mention the case of seryl-histidine, which appears to cleave DNA, proteins, and carboxyl ester (Li *et al.*, 2000). The fact that dipeptides are capable of displaying catalysis is rather interesting as it may suggest that catalysis was occuring at the early stages of life on Earth.

Sidebox 3.1

Antonio Lazcano
Facultad de Ciencias, UNAM, Mexico, D.F.
E-mail: alar@correo.unam.mx
On the abiotic synthesis of biochemical monomers and the heterotrophic theory on the origin of life
The idea that the first living systems, which resulted from a process of chemical evolution beginning with the synthesis and accumulation of organic compounds (i.e., the heterotrophic hypothesis), gained significant support in 1953. This was when Stanley L. Miller, then a graduate student working with Harold C. Urey at the

University of Chicago, achieved the first successful synthesis of organic compounds under plausible primordial conditions: with electric discharges acting for a week over a mixture of CH_4, NH_3, H_2, and H_2O; racemic mixtures of several protein amino acids were produced, as well as hydroxy acids, urea, and other organic molecules. This was followed a few years later by the work of Juan Oró, then at the University of Houston, who demonstrated in 1960 the rapid adenine synthesis by the aqueous polymerization of HCN. The potential role of HCN as a precursor in prebiotic chemistry has been supported by the discovery that the hydrolytic products of its polymers include amino acids, purines, and orotic acid, a biosynthetic precursor of uracil. A potential prebiotic route for the synthesis of cytosine in high yields is provided by the reaction of cyanoacetylene with urea, especially when the concentration of the latter is increased by simulating evaporating-pond conditions – the laboratory simulation, in fact, of Darwin's warm little pond.

The ease of formation under reducing atmospheres ($CH_4 + N_2$, $NH_3 + H_2O$, or $CO_2 + H_2 + N_2$) in one-pot reactions of amino acids, sugars, purines, and pyrimidines strongly suggests that these molecules were components of the prebiotic broth. Many other compounds would also have been present, such as urea and carboxylic acids, sugars formed by the non-enzymatic condensation of formaldehyde, a wide variety of aliphatic and aromatic hydrocarbons, alcohols, and branched and straight fatty acids, including some which are membrane-forming compounds. These reactions are effective under reducing conditions, but not if a neutral atmosphere is employed. The possibility that the prebiotic atmosphere was non-reducing ($CO_2 + N_2 + H_2O$) does not create insurmountable problems, since the primitive soup could still form, albeit from other sources.

For instance, geologically-generated hydrogen may have been available: in the presence of ferrous iron, a sulfide ion (SH^-) would have been converted to a disulfide ion (S_2^-), thereby releasing molecular hydrogen. It is also possible that the impact of iron-rich asteroids enhanced the reducing conditions, and that cometary collisions created localized environments favouring organic synthesis. Based on what is known about prebiotic chemistry and meteorite composition, if primitive Earth was non-reducing, then the organic compounds required must have been brought in by interplanetary dust particles, comets, and meteorites. A significant percentage of meteoritic amino acids and nucleobases could survive the high temperatures associated with frictional heating during atmospheric entry, and become part of the primitive broth.

It is of course surprising that amino acids can be obtained via the Strecker synthesis, purines from the condensation of HCN, pyrimidines from the reaction of cyanoacetilene with urea, and sugars from the autocatalytic condensation of formaldehyde. The synthesis of chemical constituents of contemporary organisms by non-enzymatic processes under laboratory conditions does not necessarily imply that they were either essential for the origin of life or available in the primitive environment. However, the significance of prebiotic simulation experiments is supported by the occurrence of a large array of protein and non-protein amino acids, carboxylic acids, purines, pyrimidines, polyols, hydrocarbons, and other molecules in the 4.6 billion year-old Murchison meteorite (a carbonaceous chondrite which also yields evidence of liquid water). The presence of these compounds in the meteorite makes it plausible, but does not prove, that a similar synthesis took place on primitive Earth – or is it simply a coincidence?

Other sources of organic molecules

We also know that a considerable enrichment of prebiotic moieties may have come from submarine vents and other hydrothermal sources (see, for example, Miller and Bada, 1988; Holm and Andersson, 1998; Stetter, 1998). Let's start with the 1979 discovery of deep-sea vents with black smokers, which are associated with an extraordinary abundance of the most phylogenetically primitive organisms on Earth. This ecosystem is sulphur based, and is distinct from the more familiar, photosynthetically-based ecosystem that dominates Earth's surface. Corliss *et al.* (1981) were struck by the richness of the vent biota, based on chemosynthesis, and proposed that these were the origin of life.

The sulphur-based chemistry links this work to that of Wächtershaüser, as already mentioned in the previous chapter. Let's consider again Figure 2.4, showing the formation of several important metabolites. One important metabolic cycle is the one that goes from acetic acid to pyruvate; according to the authors pyruvate is important because it occurs in various metabolic pathways, notably in the reductive citric acid cycle (see also Morowitz *et al.*, 2000). In summary, the pathway goes from the production of acetic acid through metallic-ion catalysis, to the production of three-carbon pyruvic acid by the addition of carbon; and the addition of ammonia to produce amino acids, from which peptides and proteins are finally derived. The following citation from a review by A. Lazcano (2004) summarizes the perspective in this field:

. . . primordial heterotrophs had resulted from the evolution of polymers of amino acids and other biochemical monomers produced during passage through the temperature gradient of the 350 °C vent waters to the cold ocean. Few years later, the growth and phylogenetic analysis of ribosomal RNA sequence databases led Carl Woese and his associates to conclude that the ancestors of prokaryotes had been extreme thermophiles (1979), an observation that was rapidly extrapolated by others to assume that the origin of life had taken place in sulfur-rich volcanic environments. Everything seems to fit. Transatlantic family connections led Wächtershäuser to Woese, who had been looking for an alternative to Oparin's primitive soup theory. Woese's remarks after a memorable September 1988 seminar in Woods Hole showed that he thought he had found one in Wächtershäuser's proposal.

An additional, very important source are the carbonaceous compounds coming from space, described and investigated by researchers in the field of bio-astronomy (some prefer the term "astrobiology", others "exobiology"). In recent years, bio astronomy has played a very important part in the field of the origin of life (see Zhao and Bada, 1989; Miller and Bada, 1991; Chyba and Sagan, 1992; Oró, 2002). The present amount of space dust falling on Earth amounts to *c.* 40 000 tons per year, which sounds impressive. This, spread over all Earth's surface, corresponds to only 8×10^{-9} g cm^{-2} per year (Love and Brownlee, 1993) – and the percentage of organics may be tiny (Miller, 1998; Oró, 2002). On the other hand, accumulation over billions of years can result in a considerable amount of matter. Sidebox 3.2 by Jeffrey Bada illustrates well the origin of these molecules falling on us – a taste of cosmic dust.

Sidebox 3.2

Jeffrey Bada
Scripps Institution of Oceanography
University of California at San Diego, La Jolla, CA 92093-0212
Carbon and organic compounds in the universe
The main reaction by which stars like our Sun obtain their energy is the fusion of four hydrogen atoms into a 4-He nucleus, a process called "hydrogen burning". As stars age, their hydrogen is eventually consumed and exhausted and the star then swells into a large bloated object called a Red Giant. It is in Red Giant stars that the next stage of energy and element production begins by a process called "helium-burning" in which carbon is the principal product. Carbon production involves first the fusion of two 4-He atoms to produce an unstable 8-Be nucleus, which in the interiors of hot helium-burning stars exists long enough to fuse with another 4-He atom, yielding a 12-C nucleus. All carbon present on Earth and in the Universe was once part of Red Giant stars.

Because of the tendency of carbon to form a large variety of heteroatom bonds with the elements hydrogen, oxygen, nitrogen, etc., it is not surprising that over the last 30 years astronomers have discovered a vast number of gas-phase organic molecules in space. These molecules, ranging in sizes up to more than 10 atoms, are located in regions known as interstellar clouds that contain both gaseous material as well as matter in the form of dust particles. These dust particles contain about 1% of the total mass of the interstellar medium. The gaseous components can be studied via high-resolution spectroscopy, whereas the dust particles are more difficult to characterize. Interstellar clouds can be classified as diffuse or dense, depending on their gas densities, extinctions in the visible, and temperature range. Due to these constraints, the chemistry that is taking place in these clouds is significantly different. Ion–molecule reactions are probably the most important pathways for the larger organic molecules in the gas phase of interstellar clouds. In dense clouds, dissociation–recombination reactions and neutral–neutral reactions also play a role in the formation of a variety of organic compounds. Organic compounds are thus ubiquitous in the universe. One class of molecules, polycyclic aromatic hydrocarbons (PAHs), may be the most abundant single type of molecule in space and make up more than 10% of the cosmic carbon in the interstellar medium.

Star formation and the formation of star systems with planets around them, constantly takes place in dense interstellar clouds. The material present in these clouds is incorporated into the objects that are formed during this process. Pristine or slightly altered organic matter from the cloud from which our solar-system was formed is therefore present in the most primitive objects in the solar system: comets, asteroids, and outer solar-system satellites. Pieces of asteroids (and perhaps comets) can be investigated with regards to these components through the analyses of meteorites (and eventually in samples returned from these bodies by spacecraft) in laboratories on Earth. The infall of asteroid and comet material from space may have contributed to the inventory of organic compounds on primordial Earth.

Indeed, the similarity of the molecules from space and the molecules of Earth is striking. For example, several aliphatic carboxylic acids up to C_8 were detected, (Knenvolden *et al.*, 1970), as well as alkyl phosphonates (Yuen and Knenvolden, 1973; Cooper *et al.*, 1992; Pizzarello, 1994) and also heterocyclic pyrimidines (Hayatsu *et al.*, 1975; Stocks and Schwarz, 1982).

As rich as the Murchinson and other meteorites are, comets are perhaps the most generous source of organic compounds. They are particularly rich in HCN (Miller, 1998) which appears to be very common in space and it has for this reason been investigated by several authors (Oró, 1961; Ferris and coworkers, 1973, 1974, and 1978; Matthews, 1975).

Also membrane-forming compounds may come from space (Lawless and Yuen, 1979; Deamer, 1985; Deamer *et al.*, 1994; Deamer and Pashley, 1989; McCollom *et al.*, 1999). The observation that membranogenic compounds can also be of pre-biotic origin is of particular interest, especially in view of the "membrane-first" hypothesis, or more generally for making the point that membranes and vesicles were present very early in the prebiotic scene (Deamer *et al.*, 1994). All this will be discussed further in Chapters 10 and 11.

Going back to the origin of molecules from space, it is also important to cite what has *not* been found in space: for example peptides and mononucleotides (not to mention their oligomers). On the other hand extraction and characterization of chemicals from meteorites is not easy, the work continues, and perhaps some surprises lie ahead of us.

Table 3.1 (taken from Ehrenfreund *et al.*, 2002) summarizes the relative amount of carbonaceous material coming from these different sources. Putting together

Table 3.1. *Major sources (in kg yr^{-1}) of prebiotic organic compounds in early Earth (from Ehrenfreund et al., 2002)^a*

Terrestrial sources	
UV photolysis^b	3×10^8
Electric discharge^c	3×10^7
Shocks from impacts^d	4×10^2
Hydrothermal vents^e	1×10^8
Extraterrestrial sources^f	
Interplanetary dust	2×10^8
Comets	1×10^{11}
Total	10^{11}

^{a} Assumes intermediate atmosphere, defined as $[H_2]/[CO_2] = 0.1$.

^{b} Synthesis of the Miller–Urey type.

^{c} Electric discharge may be caused, for example by lightning interacting with a volcanic discharge.

^{d} An estimate for compounds created from the interaction between infalling objects and Earth's atmosphere.

^{e} Based on present-day estimates for total organic matter in hydrothermal vent effluent.

^{f} Conservative estimate based on possible cumulative input calculated assuming a flux of 10^{22} kg of cometary material during first Ga (10^9 years) of Earth's history. If comets contain of the order of 15 wt% organic material, and if $\approx 10\%$ of this material survives, it will comprise approximately 10^{11} kg yr^{-1} average flux via comets during the first 10^9 years.

carbonaceous compounds that may have been produced from classic prebiotic chemistry, from hypothermal vents, and compounds from space, the conclusion is reached that there was enough material on Earth to start the chemistry that leads to life. Of course the distribution on Earth of these molecules was not homogenous, and one common bias is to assume that there was one place where the accumulation of material and its distribution was particularly rich and fortunate – the warm little pond of Darwin's memory. This image has given rise to the metaphor of the "prebiotic soup" – a term that has remained in the popular press, but is now rarely used in scientific literature. When it is used, it is meant broadly as the set of conditions where the wet organic chemistry of life began.

Miller's α-amino acids: why do they form?

Going back to Miller's synthesis in the flask, one question is why α-amino acids have been obtained and not, for example, β-amino acids, cyclic diketopiperazines, or some other isomers. The answer is important: α-amino acids form because they are the most stable products under the selected initial conditions. In other words the formation of those α-amino acids is under thermodynamic control. The same can be said for Oró's synthesis of adenine and other prebiotically low-molecular-weight substances formed in hypothermal vents, or found in space: certain molecules and not others form because they are thermodynamically more stable.

Is it the case that all components of our cellular life are thermodynamically stable? Of course not. We have around us many compounds in our biochemistry that are not under thermodynamic control – think of important compounds such as adenonine triphosphate (ATP), phospholipids, RNA, DNA, proteins . . . Nowadays, these compounds are formed thanks to the action of enzymes, which are often specialized for catalyzing the synthesis of products under kinetic control.

However, how could these compounds have been formed in a time without enzymes? Do we need enzymes first, in order to make biochemicals that are not under thermodynamic control? How, then, were enzymes made, since in modern times they are also produced under kinetic control? We have already seen, from the work of Wächtershaüser and colleagues, a hypothesis on how products under kinetic control might have been formed in a prebiotic scenario. However, generally speaking, this is one of the most difficult questions in the field of the origin of life: why and how have biological structures been formed that are not spontaneous, i.e. not under thermodynamic control?

The notion of "frozen accident" is often used in the literature in this connection. The term conveys the idea that something which is not thermodynamically stable may have formed by "accident" – and then been codified somehow in the living processes. Of course it is difficult to describe specific frozen accidents in terms of actual chemistry mechanisms. The notion of a frozen accident is a pictorial and fascinating metaphor, but it does not teach us anything from the operative point of view.

There is another point about the thermodynamic stability of prebiotic compounds. This is the fact that a series of thermodynamically very stable molecules seem to have been ignored in the course of prebiotic molecular evolution as building blocks of living structures. Take sugars, for example: six-membered rings have not been used for the construction of nucleic acids, where only D-ribose takes the stage. Furthermore, only two types of purine and only two of the pyrimidine bases have been utilized among the many possible nucleic acids. Actually one could make a very long list of thermodynamically stable compounds that have not been used. The general question is: "why this . . . and not that?" Or, more seriously, where does this prebiotic selection come from?

With this question we go back to some of the issues discussed in Chapter 1: contingency, and determinism, as well as intelligent design. The answer that most biochemistry students would give to this particular question (why this and not that), is that many routes were tested in prebiotic molecular evolution, and only those that "worked," namely those with some particular advantage, would have finally been chosen.

This selection argument is convincing in the case of an organism that has capability of biological competition or adaptation: those molecules and structural changes that are advantageous for say, self-reproduction, would be incorporated and become "a necessity," according to Monod's chance-and-necessity scheme. However, the selection argument is less convincing at the time of prebiotic molecular evolution, when there were still not living organisms and therefore no selective advantages.

Of course it would be easier to answer the question "why this . . . and not that" if we invoked the intelligent-design hypothesis. Conversely, one can give an explanation based on contingency, by asserting that at the time of the synthesis of, say, RNA, the contingent set of conditions were such that only D-ribose was around; and that, if in another hot spot D-furanose had been around, the reaction would have reached a dead-end. This kind of explanation is less attractive than the one based on the anthropic principle, according to which all laws of the universe have been created in a certain way, so that they could work towards life and the human species.

Is it possible to test experimentally whether the choice of a "wrong" thermodynamically stable alternative would have brought molecular evolution to a dead end? Or whether it would instead have brought about an equally probable alternative for life?

This fascinating question can indeed be tested experimentally; the approach taken by Albert Eschenmoser and his group at the Federal Institute of Technology in Zürich goes in this direction, as already mentioned (Bolli *et al.*, 1997a, b; Micura *et al.*, 1997; 1999; Eschenmoser, 1999). These authors synthesized hexopyranosyl analogs of RNA (six-membered pyranose ring instead of a five-membered furanose ring). They found that these analogs are very stable, and make strong base pairs; actually, too strong. This is not good for the function of nucleic acids, in particular replication, as at a certain point the two strands must separate from each other, and this is not possible if the base pairing is too strong. In other words, these analog structures would finally impair replication. Therefore, as Eschenmoser puts it (Eschenmoser, 2003):

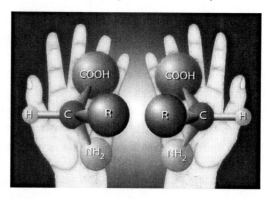

Figure 3.8 The two enantiomers of α-aminoacids. Here we follow the classic nomenclature of L- and D-aminoacids for indicating the two chiral forms. In terms of the S, R nomenclature, L-aminoacids correspond to the S absolute configuration – except for cystein, which is R.

. . . these systems could not have acted as functional competitors of RNA in Nature's choice of a genetic system, even though these six-carbon alternatives of RNA must have had a comparable chance of formation under conditions where RNA was formed.

In continuing their studies Eschenmoser and coworkers also found other structural analogs that turned out to be highly efficient informational base-pairing systems. Thus Eschenmoser arrives at this interesting conclusion (Eschenmoser, 2003):

While our experimental observations indicate that Nature, in selecting the molecular structure of her genetic system, had also other options besides RNA, the notion we naturally would be inclined to consider, namely, that RNA might be the biologically fittest of them all, remains a conjecture.[2]

This, in terms of the question "why this . . . and not that" is consistent with the view that the structures chosen by nature are not necessarily the obligatory ones. This is, of course, in keeping with the idea of contingency. The relation between contingency and determinism also forms the basis of the next issue, homochirality.

Some notes on homochirality

The question of the origin of life is usually linked to the question of the origin of molecular asymmetry, specifically why only L-aminoacids (with extremely few exceptions) are present in nature (see Figure 3.8).

---

[2] In some of his work, although in a different context, Eschenmoser is close to the question "why this . . . and not that". In his 1999 review he cites Einstein as saying: "We not only want to know *how* nature is (and *how* her transactions are carried through), but we also want to reach, if possible, a goal which may seem utopian and presumptuous, namely, to know why nature *is such and not otherwise*."

The term homochirality is generally used, and there are two levels of discussion we should consider in this regard: one is at the fundamental level of the origin of one type only of chirality – why all amino acids, and why all sugars also have only one configuration. The other point of discussion is at the macromolecular level: with chain stereoregularity, there can be variations in chirality, i.e., several types of diastereomers depending upon the distribution of chiral centers in the chain. Questions related to the stereoregularity and homochirality of chains will be discussed in the next chapter. Here some general aspects of homochirality in biomonomers will be considered.

It has already been mentioned that the origin of homochirality in nature can be viewed in terms of the controversy between determinism and contingency. Is the L form of amino acids determined by some physical law of nature; or is it a matter of chance in chemical evolution?

In fact, symmetry breaking has been seen as a result of the parity violation (Mason and Tranter, 1983; 1984; Tranter 1985a, b; Quack, 2002; Quack and Stohner, 2003a, b), according to which there should be an "ex lege" energy difference between the two enantiomers. This energy difference is minimal, of the order of 10^{-10} joule, and it is not easy to calculate it for actual molecules or macromolecules, although some attempts have been reported (Tranter *et al.*, 1992; Quack and Stohner, 2003a, b). Still, calculations are so delicate, and energy differences so small, that usually chemists and biologists prefer to neglect parity violation and similar subtle physical effects (although it is probably better to keep an open mind). In fact, in general, the existence of such laws would suggest that homochirality is more readily explained in terms of determinism.

There are other physical mechanisms proposed as sources of symmetry breaking, such as circularly polarized light, or magnetochiral anisotropy (Rikken and Raupach, 2000).

One other reason why many chemists and biologists are skeptical about parity violation and other subtle physical effects, is that the breaking of symmetry can be realized rather simply in the chemistry laboratory. According to Meir Lahav, one of the best known researchers in the field, *"breaking of symmetry is not the problem."* He means by that, that the problem is rather the propagation and amplification of chirality. In sidebox 3.3 he summarizes some of the main concepts; in particular, he considers crystals as agents of symmetry breaking (Weissbruch *et al.*, 2003).

Conceptually similar to those of Lahav are the experiments performed by Kondepudi as well as by other groups working with saturated solutions of compounds, which exist in the crystal state in two enantiomorphous forms. In particular, the primary homogeneous nucleation process of the chiral crystals of sodium chlorate and sodium bromate is very slow compared to the sequential step of crystal growth. Kondepudi *et al.* (1990) and McBride *et al.* (1991) have demonstrated that when

Sidebox 3.3

Meir Lahav

Weizmann Institute of Science, Rehovot, Israel

Crystals as agents for "symmetry breaking" and survival of homochirality in prebiotic chemistry

Crystallization and reactivity in two-dimensional (2D) and 3D crystals provide a simple route for "mirror-symmetry breaking." Of particular importance are the processes of the self assembly of non-chiral molecules or a racemate that undergo fast racemization prior to crystallization, into a single crystal or small number of enantiomorphous crystals of the same handedness. Such spontaneous asymmetric transformation processes are particularly efficient in systems where the nucleation of the crystals is a slow event in comparison to the sequential step of crystal growth (Havinga, 1954; Penzien and Schmidt, 1969; Kirstein *et al.*, 2000; Ribo *et al.*, 2001; Lauceri *et al.*, 2002; De Feyter *et al.*, 2001). The chiral crystals of quartz, which are composed from non-chiral SiO_2 molecules is an exemplary system that displays such phenomenon.

The process of "mirror-symmetry breaking" is not unique for enantiomorphous space groups, but is displayed also by non-chiral molecules that crystallize in achiral crystals at surfaces (Vaida *et al.*, 1991; Weissbuch *et al.*, 1984, 1990, 1994; Lahav and Leiserowitz, 1999; Böhringer *et al.*, 1999). Such crystals spontaneously develop pairs of chiral faces of opposite handedness. In many instances one of the faces is blocked by the surface on which the crystal resides, while the second face is exposed to the solution where the crystal might grow or absorbs chiral molecules from the environment (Weissbuch *et al.*, 1984, 1990). Both these processes are stochastic with an equal chance of the chiral crystals precipitating in either handedness, or of exposing different faces to the environment in the case of the centro-symmetric crystals. For that reason, in order to assure the survival and propagation of the chirality generated spontaneously in the first nucleation event, to the rest of the environment, it is imperative that these "mirror-symmetry breaking" processes are coupled with additional efficient steps of amplifying chirality. (Weissbuch *et al.*, 1984; Kondepudi *et al.*, 1990).

Glycine precipitates from aqueous solutions in the form of a centrosymmetric crystal, composed of alternating chiral layers of opposite handedness. These crystals are delineated at the opposite poles with two well-expressed chiral faces (01–0) and (0–10) of opposite handedness. When these crystals reside at an interface such as at the air–water or solid–water interface, only one of the chiral faces is exposed to the solution. When a single crystal of glycine attached to an interface is growing towards the aqueous solution comprising glycine and mixtures of racemic amino acid, it takes up both glycine molecules and amino acids from the aqueous food-stock; however, the amino acids are of one handedness only. The centrosymmetric crystal of the pure host has been converted into a mixed chiral crystal, composed of glycine and a mixture of amino acids of a single handedness, while the solution is enriched with the second enantiomer. While the orientation of the first crystal with regard to the solution is stochastic, as it is grown from an achiral solution, the crystals that grow in later stages are grown from a non-racemic solution. The latter exerts, via a two-step mechanism, an asymmetric induction that forces the new crystals formed at the air–water interface to assume the same orientation as the first one. Upon their growth these crystals continue to take up from the solution amino acids of the same handedness, thus resulting in the formation of a crust of glycine crystals containing amino acids of single handedness and an aqueous solution enriched with the other. (Weissbuch *et al.*, 1984, 1990).

these crystals are grown under constant stirring, once the first crystal appears, the robust stirrer fractures it into a plethora of small crystallites of the same handedness. Again, this was ascribed to a stochastic process – the first crystals were by chance of one type only, and the breaking of this first generation of crystals gave rise to many nucleation centers of the same chirality, which therefore induced selective crystallization of the same enantiomorphous form. These crystallites that are spread in the solution serve as seeds for secondary nucleation of fresh crystals of the same handedness as that of the starter. Soai and coworkers demonstrated recently that chiral crystals of $NaClO_3$ can induce asymmetry in some autocatalytic reactions (Sato *et al.*, 2004).

It was mentioned earlier that many molecules can be formed in cosmic space before arriving on Earth. What about chiral compounds? We know that amino acids are present in meteorites (Epstein *et al.*, 1987; Pizzarello *et al.*, 1994; Pizzarello and Cronin, 2000; Pizzarello and Weber, 2004). In this regard, of particular interest is the report on α-methyl amino acids, which have been found in L-enantiomeric excess in Murchison and Murray meteorites (Cronin and Pizzarello, 1997). These compounds are particularly resistant to racemization, and it is perhaps because of this that chirality has been preserved. It is not simple to assess whether these chiral exogenous compounds were the seeds for homochirality of life on Earth (Bada, 1997).

In concluding this section on chirality, let us consider one speculative question. Would life be possible with the opposite enantiomers, for example a world composed of D-amino acids? There is general agreement that this would be possible – why not, since the two enantiomers have the same energy and all the same physical properties (except for optical activity). However, there is a *caveat*. In this new chiral world, in order to have life, we would have to have the enantiomers of the sugars as well, so as to have a mirror-image world. In fact, if we had D-amino acids, but D-ribose in nucleic acids and D-glucose in polysaccharides, life as we know it would not work – as all intercrossing interactions would be diastereomeric with respect to the present ones. Think of topoisomerase, or any other enzyme acting on nucleic acids: the mirror image topoisomerase most probably would not recognize "our" current right-handed DNA. The same can be said about the interaction between proteins and membranes, containing the chirality of the glycerol ester. Thus, there appears to be a kind of complementarity between the homochiralities of the different biomonomers. If this is true, we have to find the key not only for the origin of L-amino acids, but also, simultaneously, for the D-sugars – or a causal relation between the two. Such a causal relation might be due to a primitive metabolism in which enzymes with L-amino acids make sugars, and the asymmetry of the protein molecule makes only one particular enantiomer of the sugar possible.

I have mentioned that there are two main aspects to consider on the origin of homochirality, one being the breaking of symmetry, the other the preservation and amplification of this initial inbalance. The latter will be considered further in the next chapter, which deals with macromolecules.

Concluding remarks

The first steps towards molecular complexity must have been based on spontaneous reactions – reactions that occurred because they were under thermodynamic control. As we have learned in Chapter 1, this does not mean that there is a causal chain of thermodynamic events leading to life, since a given thermodynamic output depends on the initial conditions (as is always the case in thermodynamics); these are often determined by the laws of contingency – the given temperature or pressure or concentration for that particular process. Thermodynamic control means, however, that if the same reaction is repeated under the same initial conditions, the same results are obtained – as exemplified by Miller's reaction in the famous flask under simulated reducing atmospheric conditions.

At this point, we already know that the chemistry of life is determined not only by reactions under thermodynamic control, but by a large series of reactions under kinetic control. Thermodynamic control gives products as a kind of "free lunch"; to ask the question of how and why products under kinetic control were formed, is another way of questioning the origin of life. As will be revealed later on in this book, and as already clear to most readers, macromolecular sequences are not under thermodynamic control – the primary structure of lysozyme is not as it is because of being the most stable combination of 129 amino-acid residues. In fact the aetiology of macromolecular sequences is the bottle neck of research on the origin of life. It is fine to get excited about hydrothermal vents, coupling reactions on clay, reductions by hydrogen sulfide – but with these reactions alone one does not go far. As a "Gedankenexperiment" one can offer the researchers in the origin of life all kinds of low-molecular-weight compounds in any quantity they want, including ATP and mononucleotides, lipids and amino acids, and ask them to make life – or simply to describe how life comes about. They would not know how to even start. Things would be different only if – as a continuation of the previous Gedankenexperiment – an unlimited source of enzymes and nucleic acids were to become available.

In this chapter the question of homochirality has also been considered; according to Meir Lahav breaking of symmetry is not the problem. I do not know how many scientists would agree with him, but it is certainly true that in the laboratory chiral compounds can be obtained starting from racemic mixtures – and this by simple means, without invoking subtle effects of parity violation. Of course we do not know how homochirality really evolved in nature; however, it is comforting to know that there is in principle an experimental solution to the problem.

This chapter was devoted mostly to low molecular weight compounds, and it is now the time to look at long chains, leading us to the next chapter.

Questions for the reader

1. Do you consider possible that microfossils or at least genetic material may be found in meteorites and/or comets?

2. Are you in favor of an ex-lege explanation for the onset of homochirality – or a stochastic one? Is your choice based on a sound scientific argument, or an intuition?

3. Do you have any idea how ATP might have been produced in early biochemistry? In particular: was this before or after the advent of enzymes?

4. The space mission Cassini, destined to explore the neighborhood of Saturn, will accumulate a large mass of scientific data. This mission was extremely costly. With the same money, hundreds of hospitals and schools in Africa could have been built. Where would you have invested that amount of money?

4

The bottle neck: macromolecular sequences

Introduction

Having highlighted some of the data and issues about the prebiotic chemistry of low-molecular-weight compounds, let's now turn to the functional long chains – mostly proteins and nucleic acids. The first part of this chapter is devoted to the prebiotic chemistry of biopolymers, the second part, which will necessarily be more speculative, to ideas of conceiving the very origin of macromolecular sequences.

Our biology is regulated by the catalytic power of enzymes and by the encoding power of nucleic acids. This chapter may begin with one very general question: "Why macromolecules? What is so peculiar in their great length that makes these molecules essential for life? Why didn't nature do it all using smaller peptides or smaller oligonucleotides? Why this . . . and not that?"

The question "why are enzymes macromolecules?" is an old issue in structural biochemistry, and one with which I liked to play around in my younger days (Luisi, 1979). Clearly, there are good reasons for long chains: only a long chain permits the dilution in the same string of many active residues and, simultaneously, their mutual proximity due to the forced folding; in turn, this folding and the corresponding conformational rigidity is due to the very large number of intramolecular interactions, which is only possible in long chains; the consequence of the length is an elaborate three-dimensional architecture that brings forth a particular micro-environment and reactivity of the active site; the large size is also responsible for the overall physicochemical properties, such as solubility in water or affinity to the membrane, conformational changes and cooperativity. These are all properties that can only emerge from a long chain (Luisi, 1979). Also, if enzymes are by necessity macromolecules, the responsible DNA genes must also be correspondingly long.

59

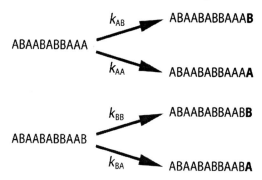

Figure 4.1 Growth of copolymers constituted by only two monomers, A and B.

Proteins and nucleic acids are copolymers

Enzymes and nucleic acids are not simply polymers, they are copolymers. This seems to be a trivial observation, and partly is; however it bears significance to an often forgotten point, that the study of the prebiotic synthesis of polypeptides and nucleic acids – and their properties – cannot ignore the general rules of copolymerization.

By way of introduction, consider the general case of the growth of a copolymer comprising only two monomers, A and B. The growing chain will either terminate with A or with B. For each of these two termini, there are two possible next steps, as illustrated in Figure 4.1: for a chain terminating with A there is either incorporation of A or B – and likewise for a chain terminating with B. Thus, in a classic linear polymerization scheme (Billmeyer, 1984; Elias, 1997) one has to consider four kinetic constants: k_{AA}, k_{BB}, k_{AB}, and k_{BA}, where k_{AA} defines the probability that an A monomer is incorporated into a chain ending with A – and likewise for the other constants. Important are the auxiliary kinetic parameters r_A and r_B, defined as $r_A = k_{AA}/k_{AB}$; and $r_B = k_{BB}/k_{AB}$. The two kinetic parameters r_A and r_B determine the tendency of the chain to assume a certain composition and sequence. This question has been extensively studied in the simple case of the radical polymerization of vinyl monomers, such as styrene, acrylonitrile, ethylene, propylene, vinyl chloride, etc. (Billmeyer, 1984; Elias, 1997), but the results are rich in information on all kinds of copolymerization. Suppose for example that both r_A and r_B are much larger than unity: then the chain with a terminal A will tend to grow in the same way (incorporating preferentially more A); and if a "mistake" – a B unit insertion – occurs, the chain will tend to further grow as BBBBB. In other words, in this case (r_A and r_B greater than one) the chain will consist of very long stretches of AAA (and long stretches of BBB) resulting in a block copolymer. Conversely, if r_A and r_B are both less than one, there will be a tendency for alternating A and B – an alternating copolymer.

These results do not depend on the absolute values of k_{AA} and k_{AB}, or k_{BA} and k_{BB}, but on their ratio, and this means that the outcome of the copolymerization cannot be predicted on the basis of the kinetic parameters of the homopolymerization of A and B alone. This also means that the copolymer product can have a composition of A and B units drastically different from the composition of the starting monomer mixture. For example, starting from a 50:50 mixture of A and B, it may happen that the resulting chains contain only a very little percentage of A. This is so, even when A has a great tendency to polymerize on its own.

In the case of vinyl monomers (styrene, acrylonitrile, acrlylamide, isobutylene, etc.) copolymerization is generally spontaneous; however, the reaction products are determined by the kinetic constants – a case of interplay between thermodynamic and kinetic factors.

In the case of proteins or nucleic acids we do not have two, but several comonomers; furthermore we are not dealing with the simple case of radical polymerization, but with the more complex polycondensation. Very little is known about the kinetics of the copolymerization of polycondensates – for example analysis of r_A and r_B has not been done systematically for amino acids. However, a few general points can still be made on the basis of the general principles of copolymerization. One has been already mentioned: that the initial composition of amino acids in the "prebiotic soup" may not correspond to the amino-acid composition in the chain. Thus, the fact that one given amino acid has a very small frequency of occurrence in protein chains may not necessarily mean that this amino acid was not present under prebiotic conditions; the low frequency in the chains can simply be the result of the kinetics of polycondensation. Conversely, the presence of preferred residues or short sequences in protein chains might be due to the interplay of kinetic parameters, and have little to do with the initial biological constraints.

Kinetic constants and kinetic parameters measure probabilities, and of course in a real copolymerization process one eventually should consider velocities – namely take into account the actual concentration of A and B. Then prediction of composition and preferential sequences becomes even more complicated (Billmeyer, 1984; Elias, 1997).

More generally, these few considerations indicate the complexity of the polymerization of biomonomers. In particular, when considering the aetiology of polypeptide sequences, one cannot assume that all amino acids are equivalent to each other in their rate of incorporation; a given sequence is the result of an extremely large number of kinetic and thermodynamic factors. This is so even for relatively short oligopeptides.

All this is perhaps not very encouraging, but it may serve to make us view the experimental data presented in the literature more critically. Also, let us not brush off the question of the polymerization of polypeptides with the arguments that they are just derived from nucleic acids: of course all the above-mentioned difficulties also hold for the polycondensation of nucleic acids.

To the theoretical difficulties mentioned above, one should add the operational ones. Those chemists who are involved with the synthesis of oligopeptides are plagued by the fact that peptides tend to be insoluble. Also, it is well known that in aqueous solution the reaction leading to chain condensation starting from amino acids is thermodynamically unfavorable, even when starting from the corresponding amides. The same thermodynamic difficulty holds for the condensation of mononucleotides into polynucleotides (Ferris, 1998). Thus, an energy input is necessary in order to make chains. In chemical terms, this usually means that the terminal bonds must be activated.

Chemical activation is indeed the weak point of the prebiotic chemistry of polycondensation. In principle, this should be a "prebiotic activation," namely a kind of spontaneous reaction under prebiotic conditions. Several more or less "friendly" activation methods have been used in the field, and most of them cannot, reasonably, be called prebiotic. On the other hand, the chemist must start working with some tool at hand. Let us now take a few examples from the literature on the prebiotic synthesis of biopolymers.

The quest for macromolecular sequences

Let us consider first homopolypeptides, i.e., polymers of only one type of amino acid. Since it is difficult to form long chains in water, the method of choice has for a long time been to carry out chemical reactions on clay surfaces with the idea of obtaining mineral catalysis as well (Theng, 1974; Kessaissia *et al.*, 1980; Bujdak *et al.*, 1994; 1995; Schwendiger and Rode, 1992). The competition with water hydrolysis is of course the main problem, and one can infer (Ferris, 2002) that in the presence of water, only polymers capable of undergoing slow hydrolysis can be expected to grow into long polymeric chains. In fact, relatively long homopolymers containing glycine or glutamic acid were prepared (Ferris *et al.*, 1996; Radzicka and Wolfenden, 1996; Orgel, 1998). The advantages and disadvantages of this procedure have been discussed by Alan Schwarz (1998).

On the basis of what was said earlier about copolymerization theory, these results cannot be generalized for the case of co-oligopeptides.

On the subject of montmorillonite clay, the quite interesting reaction reported by Paechthorowitz and Eirich (1988) should be mentioned. The reaction sequence involves the coupling of an amino acid to form a polypeptide–nucleotide compound, as shown in Figure 4.2.

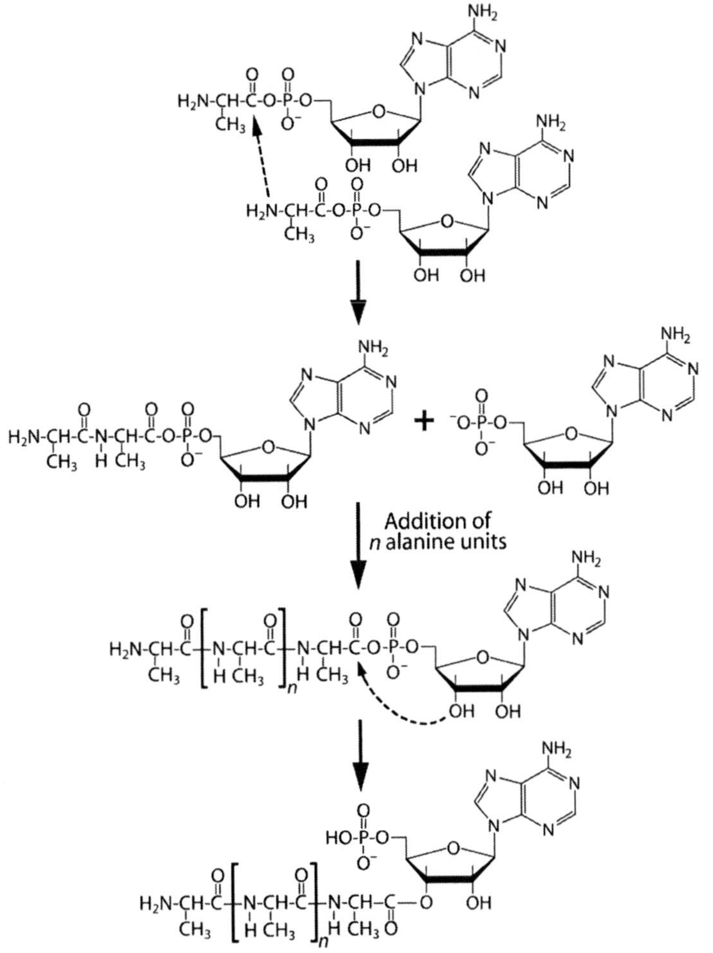

Figure 4.2 From top to bottom, a reaction sequence in which an amino acid (alanine) reacts with a nucleotide to form a polypeptide–nucleotide compound. Although reactions of this kind proceed on montmorillonite clay in the presence of water, their precise mechanisms (at the curved arrow, for example) have yet to be determined. (Adapted from Ferris, 2002.)

In particular, the aminoacyl phosphate derivatives of 5′-AMP could condense to a polypeptide containing up to 56 residues. This is certainly a very elegant and impressive reaction, but its relevance from the point of view of prebiotic chemistry is doubtful since, as already mentioned, it is not obvious whether and to what extent this reaction can be extended to co-oligopeptides. Also, of course, AMP cannot be considered (thus far) a prebiotic molecule. Here again we encounter the old problem of the RNA world, the fact that the prebiotic synthesis of mononucleotides is not understood. There is no doubt that if a robust and credible prebiotic route to AMP was found, the reaction shown in Figure 4.2 (if extended to mixtures of amino acids) might be of great importance for building macromolecular sequences. Even under these conditions, one problem would remain: the order of amino acids in the chain. How could the reaction be regulated so as to produce one given sequence with many identical copies, instead of a chaotic mixture of many different chains?

Staying with the simpler case of homopolypeptides, there is a particularly interesting report by Tsukahara *et al.* (2002) on prebiotic oligomerization on or inside lipid vesicles in a hydrothermal environment; oligopeptides up to heptaglycine were formed even in the absence of condensing agents. This would imply that the environmental conditions (perhaps the local high concentration reached in vesicles) were instrumental in favoring the peptide coupling even without chemical activation.

Let us consider now the question of the synthesis of polypeptides containing different residues in the chains (co-oligopeptides). The Merrifield procedure is commonly used in order to obtain very long co-oligopeptides with a specific sequence, but this method is usually not regarded as prebiotic. In fact, we simply do not know how to make long polypeptides by prebiotic means. I know that the late Sidney Fox would not have agreed with this statement, and it is important to mention his work (Fox and Dose, 1972; 1977; Fox, 1988). On heating mixtures of amino acids (containing a ten-fold excess of residues with reactive side chains, such as glutamic acid, aspartic acid, or lysine) at 180 °C for a few hours, the authors showed that "proteinoids" could be formed, i.e., bodies containing polymerized amino acids. However this work has never been repeated with a thorough characterization of the reaction products, and, rightly or not, is not considered a reliable method. It was, however, reported that when using amides, the presence of clay increases the yields during repeated drying and heating, and Ito and coworkers (1990) reported a substantial array of polypeptides prepared in this way. However, generally, one can in good faith accept the assertion that no reliable method is known to produce high-molecular-weight co-polypeptides under prebiotic conditions.

Still with the prebiotic scenario, the conditions developed by Limtrakul *et al.* (1985) are interesting: as a consequence of evaporation, high local concentrations may have arisen, and under such conditions the incompletely hydrated metal ions may activate a dehydration leading to peptide condensation. From this, the technique of the salt induced peptide condensation (SIPC) has been developed (Oie *et al.*, 1983; Suwannachst and Rode, 1999; Rode *et al.*, 1999).

There is some claim that long polypeptide chains may derive not so much from the condensation of amino acids, but from the polymerization of HC≡N followed by simple prebiotic chemistry to mould the side chains (Matthews, 1975); however, this theory has not yet found great support in the field.

Finally the condensation of *N*-carboxyanhydrides (NCAs) (also known as Leuch's anhydrides) deserves mention. The relevance of this reaction, already illustrated in Figure 3.7, lies in the fact that NCA-amino-acid derivatives are supposed to be prebiotic compounds (Taillades *et al.*, 1999). As pointed out by Ferris (2002), this synthetic route has several advantages with respect to other routes. In fact, the synthesis can occur in aqueous solution, since the polymerization is faster than the hydrolysis rate; there is no racemization and the synthesis is specific for α-amino acids. Oligomers of up to ten can be obtained in one single step; this is of course also the limit of the method, as it appears impossible to reach significantly higher degrees of polymerization. However, decamers can be seen as interesting precursors for prebiotic-fragment condensation, as will be argued later in this chapter.

A new interesting development has been offered by Orgel and coworkers (Leman *et al.*, 2004): they showed that carbonyl sulfide (COS), a simple volcanic gas, brings about the formation of peptides from amino acids under mild conditions in aqueous solution, and in yields approaching 80% in minutes to hours at room temperature. Dipeptides and tripeptides were thus obtained, but in this case too the answer to the question of long chains with a regulated order of sequence remains elusive.

Looking at the few illustrative data presented here on the polycondensation of amino acids into polypeptide chains, one can conclude that a variety of interesting reactions have been proposed and studied, some of potential interest in the prebiotic scenario. However, the question of how to make long and specific sequences in a way relevant to the origin of life is still open. For example, there is very little in the literature for the synthesis and *characterization* of copolypeptide chains containing, say, 30 residues, obtained under alleged prebiotic conditions, (up to the spring of 2006). Generally then, the question "why this ... and not that?" – namely why a particular polypeptide sequence was formed, and not a different one, is still unanswered.

What about polynucleotides?

Having considered polypeptides, let's now turn our attention to polynucleotides. In this field, as already mentioned, the problem is still that researchers are still actively investigating possible routes to the synthesis of mononucleotides.

Remarkable work has been carried out by James Ferris and coworkers, who showed among other things that a common clay mineral, montmorillonite, permits the catalytic oligomerization of activated mononucleotides such as adenosine-5'-phosphorimidazolide (Ferris and Ertem, 1992; 1993; Ding *et al.*, 1996). The reaction scheme for the oligonucleotide elongation is shown in Figure 4.3. Eventually high

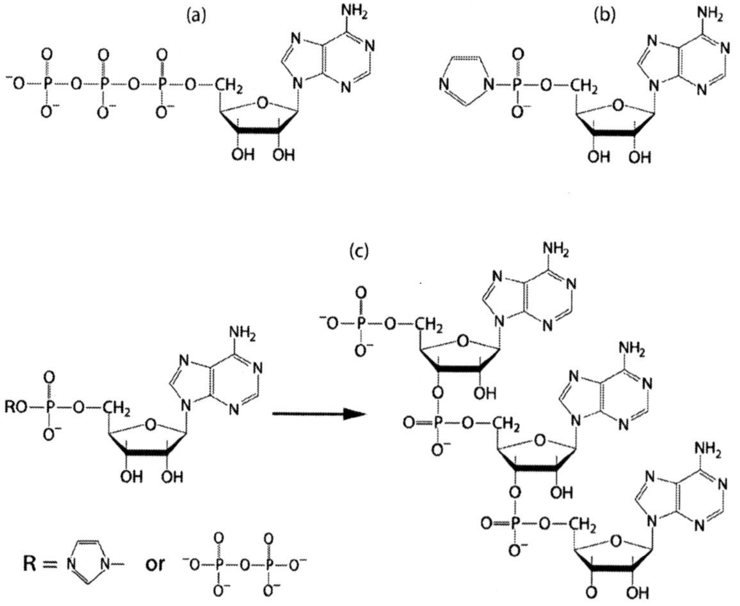

Figure 4.3 The synthesis of an oligonucleotide from an activated mononucleotide. (a) Adenonine triphosphate (ATP), the substrate of enzymatic nucleic-acid synthesis. (b) An imidazolide of a nucleotide of the kind used in many non-enzymatic template-directed reactions. (c) The synthetic reaction leading to the formation of a trinucleotide. (Modified from Orgel, 2002.)

degrees of polymerization were reached (Ferris *et al.*, 1996; Ferris 1998), partly with the stereospecific 3′–5′ linkage along the chain.

Also interesting is the work by Franchi and Gallori (Franchi *et al.*, 2003; Franchi and Gallori, 2004), who showed that once nucleic acids are adsorbed on clay, they are much more resistant to irradiation, temperature shock, and other extreme conditions.

In a chapter on polymerization of mononucleotides, the field of template-directed oligonucleotide synthesis, as pioneered by L. Orgel and his group, should also be mentioned. The starting consideration is that a given sequence of nucleotide bases is in principle capable of aligning the complementary sequence. In fact, it has been observed that monomer derivatives of purine nucleotides align spontaneously on complementary polymers of pyrimidine nucleotides (Howard *et al.*, 1966; Huang and Ts'o, 1966). Leslie Orgel, in his recent review (Orgel, 2002), gives additional examples; he makes the point that all these data suggest:

. . . that the formation and copying of long, single stranded, RNA oligomers could have occurred spontaneously, given a pool of activated monomers.

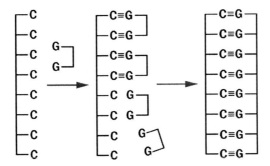

Figure 4.4 Orgel's template-directed oligomerization. The main process in template-directed oligomerization of mononucleotides actually occurs at the dimer or trimer level. When an activated mononucleotide is allowed to react in the presence of a complementary polynucleotide template, dimers are first formed in a non-catalyzed phase of the reaction; they then bind to the template and are ligated and extended. (Adapted from Schwarz, 1998.)

$$C - C - G - C - C$$
$$pG \cdots G \cdots C \cdots G \cdots G$$

Scheme 4.1.

Orgel's pioneering work has been instrumental for the maturation of a series of important concepts in the field. The general reaction scheme is illustrated in Figure 4.4.

This scheme can indeed be implemented by chemistry: thus, the pentamer pGGCGG, for example, was obtained from a mixture of 5′-(2-methyl)-phosphorimidazolide derivatives of guanosine and cytidine in the presence of the complementary CCGCC as a template (Inoue and Orgel, 1983): see Scheme 4.1. Later, even longer templates were successfully used (Orgel, 1992); for a review of these, and other work, see Orgel, 1995.

Recent work from Christoff Biebricher's group showed that a very efficient template-directed polymerization of adenine – with a degree of polymerization of up to 400 – could be achieved based on a poly(U) template (Trinks *et al.*, 2003).

Self-replication of long nucleotide sequences has not been achieved based on template-directed synthesis, but Gunter von Kiedrowski and colleagues were successful in 1986 with a slightly different approach (von Kiedrowski, 1986). This point will be readdressed in Chapter 7. The elongation of oligonucleotides has been studied in a different context by Eschenmoser's group (Bolli *et al.*, 1997a, b), with emphasis on the retention and origin of homochirality, and will be considered later on in this chapter.

In conclusion, much elegant work has been done starting from activated mononucleotides. However, the prebiotic synthesis of a specific macromolecular sequence does not seem to be at hand, giving us the same problem we have with polypeptide sequences. Since there is no ascertained prebiotic pathway to their synthesis, it may be useful to try to conceive some working hypothesis. In order to do that, I would first like to consider a preliminary question about the proteins we have on our Earth: "Why these proteins . . . and not other ones?". Discussing this question can in fact give us some clue as to how orderly sequences might have originated.

A grain of sand in the Sahara

This is indeed a central question in our world of proteins. How have they been selected out? There is a well-known arithmetic at the basis of this question, (see for example De Duve, 2002) which says that for a polypeptide chain with 100 residues, 20^{100} different chains are in principle possible: a number so large that it does not convey any physical meaning. In order to grasp it somewhat, consider that the proteins existing on our planet are of the order of a few thousand billions, let us say around 10^{13} (and with all isomers and mutations we may arrive at a few orders of magnitude more). This sounds like a large number. However, the ratio between the possible (say 20^{100}) and the actual chains (say 10^{15}) corresponds approximately to the ratio between the radius of the universe and the radius of a hydrogen atom! Or, to use another analogy, nearer to our experience, a ratio many orders of magnitude greater than the ratio between all the grains of sand in the vast Sahara (see Figure 4.5) and a single grain.

The space outside "our atom", or our grain of sand, is the space of the "never-born proteins", the proteins that are not with us – either because they didn't have the chance to be formed, or because they "came" and were then obliterated. This arithmetic, although trivial, bears an important message: in order to reproduce our proteins we would have to hit the target of that particular grain of sand in the whole Sahara.

Christian De Duve, in order to avoid this "sequence paradox" (De Duve, 2002), assumes that all started with short polypeptides – and this is in fact reasonable. However, the theoretically possible total number of long chains does not change if you start with short peptides instead of amino acids. The only way to limit the final number of possible chains would be to assume, for example, that peptide synthesis started only under a particular set of conditions of composition and concentration, thus bringing contingency into the picture. As a corollary, then, this set of proteins born as a product of contingency would have been the one that happened to start life. Probably there is no way of eliminating contingency from the aetiology of our set of proteins.

Figure 4.5 The ratio between the theoretical number of possible proteins and their actual number is many orders of magnitude greater than the ratio between all sand of the vast Sahara and a single grain of sand. (Picture kindly provided by Giuseppe Carpaneto.)

The other objection to the numerical meaning suggested by Figure 4.5 is that the maximum number of proteins is much smaller because a great number of chain configurations are prohibited for energetic reasons. This is reasonable. Let us then assume that 99.9999% of theoretically possible protein chains cannot exist because of energy reasons. This would leave only one protein out of one million, reducing the number of never-born proteins from, say, 10^{60} to 10^{54}. Not a big deal.

Of course one could also assume that the total number of energetically allowed proteins is extremely small, no larger than, say, 10^{10}. This cannot be excluded a priori, but is tantamount to saying that there is something very special about "our" proteins, namely that they are energetically special. Whether or not this is so can be checked experimentally as will be seen later in a research project aimed at this target.

The assumption that "our" proteins have something special from the energetic point of view, would correspond to a strict deterministic view that claims that the pathway leading to our proteins was determined, that there was no other possible route. Someone adhering strictly to a biochemical anthropic principle might even say that these proteins are the way they are in order to allow life and the development of mankind on Earth. The contingency view would recite instead the following: if our proteins or nucleic acids have no special properties from the point of view of thermodynamics, then *run the tape again* and a different "grain of sand" might be produced – one that perhaps would not have supported life.

Some may say at this point that proteins derive in any case from nucleic-acid templates – perhaps through a primitive genetic code. However, this is really no argument – it merely shifts the problem of the etiology of peptide chains to etiology of oligonucleotide chains, all arithmetic problems remaining more or less the same.

The "never-born proteins"

To look at the vast Sahara desert is challenging from the point of view of biochemical research. What do all these "never-born proteins" (NBP) look like? Are they only trivial variations of the proteins we already know, or – since there is such an immensity of them out there – would it be possible that some of them possess unknown structure and properties? Have they not been produced simply and solely because of lack of time and bad luck – or due to the concomitance of some unknown and more subtle reasons? (Of course we have many still unknown proteins on our Earth, but clearly the question of the never-born proteins has a quite different flavor, as they were never selected).

A simpler question related to structure concerns the folding. We know that "our" proteins display a biological function only when they assume a specific conformation. Let us assume to be able to make a large library of NBPs: what would be the frequency of folding – namely which percentage of them would assume a stable tertiary conformation?

This question can be tackled with a concrete research project, which started recently in my group. The basic idea is to make a large number of NBPs 50 residues long. The maximum theoretical number of such chains is 20^{50}, and by using the well-known technique of phage display, gene libraries containing approximately 10^9 clones can be produced.

Of course it is not difficult to make novel DNA sequences which are 150 base pairs long: by randomly designing a gene that long, the probability of hitting one already existing in our living world is given approximately by the "Sahara ratio" of Figure 4.5. The randomly designed genes for making the library were totally random except for an inserted tripeptide, which is the substrate for thrombin (see Figure 4.6).

In principle, it is not fair, or course, to approach a problem of prebiotic chemistry by using sophisticated techniques of present-day molecular biology, such as phage display. However in this case the particular research question was not the origin of life, but rather the question: given a vast library of random polypeptide chains, what is the folding frequency? The criterion utilized for determining whether a protein is folded or not was based on resistance to the hydrolytic power of proteases, with

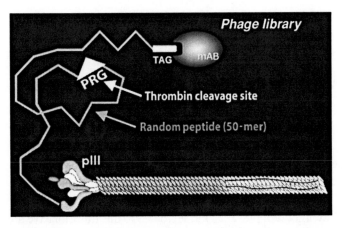

Figure 4.6 The phage display scheme for the production of never-born proteins. Proline-arginine-glycine (PRG), is a substrate for thrombin, and TAG is the antibody target. (Modified from Chiarabelli *et al.*, in press.)

the reasoning that folded chains are much more resistant than unfolded ones to proteolitic digestion. It is not of course an absolutely sound criterion, but it can serve as a first qualitative test. Using the resistance to the action of thrombin as a first approximation for the folding, and analyzing a randomly taken sample of approximately 80 clones, it was found that almost 20% of the chains were resistant to the action of thrombin (Chiarabelli *et al.*, unpublished data).

If these preliminary data hold true and a large fraction of the never-born proteins tend to assume a stable tertiary folding, then one could assert that folding is not a particular property of "our proteins", but a general one. Also, the presence of a huge number of folded chains is likely to give a high probability of a consequent specificity, for example in the binding or even in the catalysis. Seen in this light, contingency is associated with a relatively high probability of folding, which is good.

There is another general lesson that we can learn from these experiments. If "our" proteins are the product of contingency, most probably the pathway to their prebiotic synthesis cannot be reproduced in the laboratory. This is indeed the bottle neck in the bottom-up approach to the origin of life.

What can we do then in our laboratory? We can resort to the observation of Eschemoser and Kisakürek (1996) already mentioned in Chapter 1, according to which what is important is to show that the pathway is possible – in this particular case, that it is possible to construct many identical copies of a given long polypeptide or nucleotide sequence, capable of binding and catalysis.

Generally, I believe that tackling experimentally the question "why this . . . and not that" – to try to understand namely why the products of our nature are in one form and not another – may be very instructive for the field of the origin of life and chemical evolution.

A model for the aetiology of macromolecular sequences – and a testable one

Let us go back to the problem of the formation of specific macromolecular sequences. The core of the problem lies in the fact that the synthesis of such copolymeric sequences, like lysozyme or t-RNA^{phe}, is not under thermodynamic control. How can we then conceive their formation under prebiotic conditions, i.e., in a time where only spontaneous reactions were possible?

In the following I will present a model to tackle this question. We are at the level of speculations, but we will use some known facts. For example, let's start from the condensation of NCA-anhydrides, a reaction which, as we have seen in the previous chapter, is considered prebiotic. In this way, oligopeptides up to (say) a length of ten can be built, possibly under thermodynamic control, but starting from a definite set of conditions (amino-acid composition in the starting mixture, pH, salinity, etc.). We can assume, and this is actually quite reasonable, that in this way copious libraries of different decapeptides have been formed, each in a significant concentration.

Once we have this library, fragment-condensation procedures might link the oligopeptides to one another so as to build longer chains. Prebiotic methods for fragment condensation are not really known. Kent's synthesis comes close to the target (Kent, 1999) but is restricted to particular chemical structures and cannot be seen as a generalized prebiotic method.

Since the chemistry for fragment condensation is not known, we need some assumptions. One major assumption is now that the prebiotic fragment polycondensation was catalyzed by peptides originating from the NCA-condensation reaction itself.

The idea is not new. The hypothesis that long protein chains may be derived from the condensation of shorter peptides has been around for some time; and it is also accepted that shorter peptides may have some catalytic activity. One example is the catalytic activity of prebiotic histidyl-histidine (Shen *et al.*, 1990a; b). Also glycine and diglycine have been postulated as possible catalytic factors in the prebiotic evolution of peptides (Plankensteiner *et al.*, 2002). White and Erickson (Erickson *et al.*, 1980) reported the catalysis of peptide bonds due to the action of histidyl-histidine in a fluctuating clay environment, adding that this might indeed be a model for a primitive prebiotic synthesis by a "protoenzyme." And we have already mentioned the work of Yufen Zhao's group with seryl-histidine and related oligopeptides, which appear to cleave DNA and proteins (Li *et al.*, 2000).

Fragment condensation of peptides corresponds to a reverse protease reaction – peptide synthesis instead of cleavage – and this is well known in the literature as well. In fact, proteases have been used extensively for peptide coupling (Jakubke *et al.*, 1985; 1996; Jakubke, 1987; 1995; Luisi *et al.*, 1977 b). This work has shown that even small proteins can be synthesized by block-wise enzymatic coupling (see also Kullmann, 1987, and, for some more recent developments, Celovsky and Bordusa, 2000).

The reverse reaction takes place of course under particular conditions, and various methods have been conceived to do this. For example a reaction may take place on reducing water concentration by freezing the aqueous reaction mixtures (Hansler and Jakubke, 1996) as well as by carrying out syntheses in organic solvent-free media containing a minimum water content (Halling *et al.*, 1995; Eichhorn *et al.*, 1997); or using clay support (Bujdak *et al.*, 1995; Zamarev *et al.*, 1997; Rode *et al.*, 1999); or by making the product insoluble, so that there is phase separation that shifts the equilibrium towards the product (Luisi *et al.*, 1977b, Anderson and Luisi, 1979; Lüthi and Luisi, 1984). The scenario of dry and wet cycles, invoked by several authors in the prebiotic field, may be another way of bringing forth the stepwise elongation (see, for example, Saetia *et al.*, 1993).

Let us see where this scheme may bring us, and for this purpose refer to Figure 4.7. This shows a small library of oligopeptides, where one (or more) of these products, indicated in the figure with an asterisk, is endowed with peptidase activity. These peptides then, under the reaction conditions, can catalyze fragment condensation in a specific way – for example of those peptides with a terminal aromatic residue. Following the first condensation step, the catalytic center is inserted into a longer polypeptide, with retention of the catalytic activity.

We can concretize these speculations with some figures. Let us assume that the initial mixture of NCA-activated amino acids affords ten different 10-peptides, one with proteolytic activity; and that three out of these ten are substrates for the proteolytic peptide (e.g., with an aromatic group). Fragment condensation would give rise ideally to nine different 20-peptides bearing again an aromatic group at the end of the chain; three of them, ideally, are still catalytically active. The elongation can then proceed further, giving rise ideally to up to 81 polypeptides that are 40 residues long. Those, in principle, are capable of further elongation . . .

Of course one can also imagine (and this is again not too far-fetched) that the contingent environmental conditions – pH, temperature, salinity etc, – would eliminate a good deal of these products (because they are, for example, insoluble under the given conditions). As suggested by Figure 4.7, these unfit products are eliminated from further elongation steps. In this way, the postulated mechanism of chain elongation can be seen as being attended by a kind of evolutionary environmental pressure. In fact, there is in this scheme an interplay between contingency of the

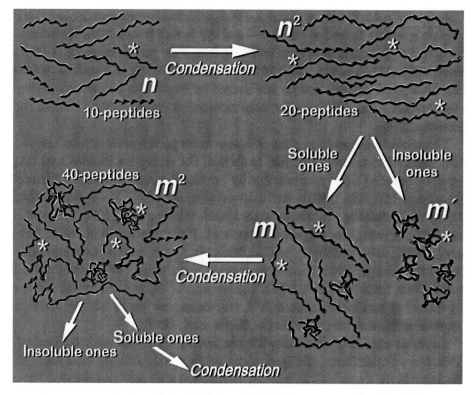

Figure 4.7 Fragment condensation of short, prebiotically formed, oligopeptides. The asterisk indicates the catalytically active peptide, which can induce the fragment condensation by reverse peptide hydrolysis; n peptides (e.g., ten residues long) react with each other to build ideally n^2 20-peptides, and of these, m react further (to build ideally m^2 40-peptides, of course in practice all possible mixtures may be present), and m' are eliminated because of being unusable (e.g., insoluble) under the contingent conditions – and so on.

environmental conditions and the obligatory pathway due to the enzyme specificity and to the thermodynamic control.

Is this scheme realistic? It is difficult to say, but some characteristics of this hypothetical general mechanism are worth noting. The fact that the reaction products are under thermodynamic control gives an initial flavor of determinism. However the initial composition of the peptides would be due to contingency. Also the rest of the mechanism is based on an interplay between specificity (determinism) and contingency. The specificity of the catalytic reaction would provide a certain degree of chemical selection, for example due to the terminal aromatic reactive amino acid. In addition, as mentioned above, some of the peptides may not be able to react because they are insoluble under the contingent set of environmental conditions (such as temperature, pH, salinity etc.) or because they may form aggregates.

Another characteristic of such a mechanism is autocatalysis, as the number of cat- alytically active species increases with each condensation step (assuming that the catalytic activity is maintained in the elongation process).

Still regarding figures: how would the mechanism shown in Figure 4.7 work with real quantities? Suppose that each of the three initial substrate oligopeptides in our theoretical scheme were present at a concentration of 3 mg ml^{-1}. On the basis of an average yield of 10% at each condensation step, concentrations of 0.6 mg ml^{-1} for the 20-mers, and 0.12 mg ml^{-1} for the 40-mers, and of 0.024 mg ml^{-1} for the 80-mers would be obtained. This last figure corresponds to a molar concentration of these macromolecules of about 3 μM.

The elongation step under a simulated environmental pressure is possible, as a never-born protein with 43 residues was obtained, although not by catalytic frag- ment condensation but by the Merrifield method (Chessari *et al.*, unpublished data).

In fact, the good thing about this scheme is that it can be experimentally tested. It is possible to ascertain whether and to what extent the NCA-polycondensation products are under thermodynamic control (giving rise to the same compounds in the same amount each time the reaction is carried out); also, whether, once formed, the oligopeptides can be coupled together by the action of some real proteolytic enzymes; finally one can also test whether the products of the NCA mixture pro- duces peptides with proteolitic activity. In fact, this kind of work is being carried out in our laboratory in collaboration with Peter Straweski. It is tempting to speculate that such a mechanism might have been at the basis of the origin of macromolecular sequences.

Fantasy can play further with the scenario depicted in Figure 4.7. For example, the mechanism can in principle be extended to oligonucleotides. One needs to assume here that NCA-formed oligopeptides were capable of catalyzing the coupling of nucleotides; stretching this picture a little further, one may also assume that the formation of the mononucleotides from the three moieties might have been due to peptide catalysis. It is in fact already known that hexanucleotides containing guanine and cytosine can be condensed using water-soluble carbodiimide (Kawamura and Kamoto, 2000). The problem is that in this case we need three or four different reverse hydrolase activities.

Having indulged in fantasy, let me add a little more of it – by bringing vesicles into the picture. Vesicles might to a certain extent have facilitated intermolecular reaction mechanisms and also, through vesicle fusion, enrichment of the initial metabolism. This can be taken a little further permitting the chimerical illustration in Figure 4.8, showing the fusion between peptide- and nucleotide-containing vesicles. This is somehow reminiscent of an old idea by Dyson, who in his classic book on the origins of life (Dyson, 1985) proposed the notion of a double origin of life, one for proteins and independently one for nucleic acids, two worlds that eventually

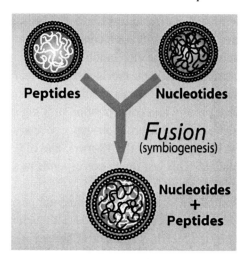

Figure 4.8 The hypothetical prebiotic fusion between nucleotide- and peptide-containing vesicles.

may have fused with each other so as to give the real beginning of life. It would be practically a kind of symbiogenesis, a notion dear to Lyn Margulis (Margulis, 1993), but this time at the level of prebiotic life.

What is good about the field of the origin of life, is that there is so much that we do not know, that speculations are allowed and actually welcome; of course there are good and bad ones. The good speculations are those that not only solve on paper a still unsolved problem, but can be tested experimentally; and in this way rise to the level of working hypotheses.

However, we ought now to leave speculations and move back into a more descriptive reality.

Homochirality in chains

It was mentioned earlier that homochirality gives the question "why this and not that" a particular meaning in the case of macromolecules. Why homochiral macromolecules? Assume for a moment that both L and D forms of the natural α-amino acids are present. Then, a protein with 50 residues would in principle exist in all its 2^{50} diastereoisomeric forms. By utilizing only one enantiomer, nature reduces this number 2^{50} to *one*! What a trick!

These arguments suggest that homochirality was most probably already implemented in the early stages of the origin of life, as it would have been very difficult to reach the high order and selectivity of the first self-reproducing cells if all possible diastereomeric macromolecules had been around.

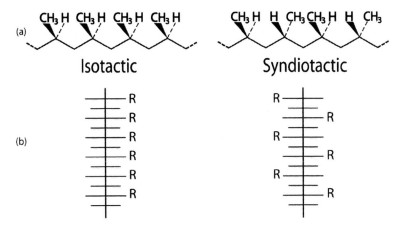

Figure 4.9 Stereoregular vinyl polymers in (a) the ideal zig-zag conformation, and (b) in the Fischer projection. If the termini of the chain are chemically different, all tertiary C atoms assume the same configuration and each isotactic chain becomes chiral. When the two chain termini are identical, each chain is superimposable with its mirror image. When R = methyl, we have polypropylene.

Natural biopolymers are homochiral, as they are constituted by L-amino acids and by D-sugars. In this sense, they are stereoregular, to use the nomenclature of classic macromolecular chemistry. Occasionally, also in the field of peptide chemistry, the term "isotactic" is used to describe polypeptides formed by only one type of monomer chirality. The term isotactic was actually coined for stereoregular vinyl polymers – polypropylene in particular – to describe a situation in which the tertiary carbon atoms of the monomer units all have the same absolute configuration (when the chain has different chain termini) as illustrated in Figure 4.9. Another form of stereoregularity is the one in which there is a regular alternation of monomer-unit configurations, which forms a syndiotactic polymer. In the case of a random statistical distribution, one can have "atactic" chains (no stereo-order along the chain); or chains containing long sequences of one chirality alternating with long sequences of the opposite chirality (block homochiral sequences).

The two L and D monomers can be viewed as the two comonomers, and then stereoregularity is the particular outcome of a copolymerization scheme as formally described by Figure 4.1. Although stereoregular vinyl polymers are not directly related to the world of biopolymers, they can teach us a couple of points of general importance.

One of the motivations for the Nobel prize being awarded to Giulio Natta (together with Karl Ziegler) was that, by building isotactic polypropylene (PP) macromolecules, he had broken the monopoly of nature on stereoregularity: see Figure 4.9. Another analogy with the world of the natural biopolymers lies in the fact that isotactic PP chains, because of their regular structure, are able to assume helical conformations in the crystalline state. Of course left-handed and right-handed helices are mirror images of one another and are present in the same amount. This is not the case for the α-helix of natural polypeptides, which is present only in the right-handed form. The reason for this difference is obvious: in the case of proteins only L-amino acids are present, and the right-handed and the left-handed helix, both with L-amino acids, are diastereomers – not mirror images as in the case of isotactic polypropylene – and therefore have different energies.

To mention the monopoly of nature is proper, as the biopolymers of life display a rich array of stereoregularity: the L configuration of amino acids, and in the case of nucleic acids the D configuration of ribose, and the 3′–5′ link of the nucleotides; for polysaccharides, the same absolute configuration of D-glucose, and the 1–4 glycosidic linkage. Also worthy of mention are natural polyisoprene, kautschuck and gutta-percha, which have regular successions of *cis* double bonds and *trans* double bonds, respectively.

As already mentioned, this high stereochemical order is a case of kinetic control, as all these polymers are the product of highly specialized enzymes. Still, chain stereoregularity in polypeptides and/or polynucleotides has some additional features of interest for the field of the origin of life. This is covered in the next sections.

Chain chirality and chain growth

In particular, I would like to deal with some aspects of the relation between chain growth and stereoregularity. Recognizing that functional chains must be stereoregular, and supposing an original prebiotic world inhabited by racemic amino acids, the question may be asked: in a prebiotic world without the stereoregulatory power of enzymes, is it possible to arrive at some kind of homochirality?

Let us consider first a situation in which the starting mixture already contains a very small enantiomeric excess of one amino acid over the other, as that observed in meteorites. Let us assume an enantiomeric excess (e. e.) of 0.2 – this is the ratio $(D - L)/(D + L)$ – corresponding to a composition of 60% D and 40% L units. Let us follow the growth of this polymer assuming – in the simplest case – an ideal copolymerization of L and D units, namely no difference in the rate of incorporation in the growing chain. In particular, let us calculate the ratio D_n/L_n where n is the length of the chain. In the monomer mixture, this ratio is, in our example, 1.2. Now notice that this ratio increases with n according to a trivial binomial distribution, as $(D_n/L_n)^n$. The enantiomeric excess can then be calculated, expressed by the ratio $(L_n - D_n)/(L_n + D_n)$ as a function of n, and the conclusion is reached that even at $n = 10$, the theoretical enantiomeric excess is $c.$ 100%. In other words, there

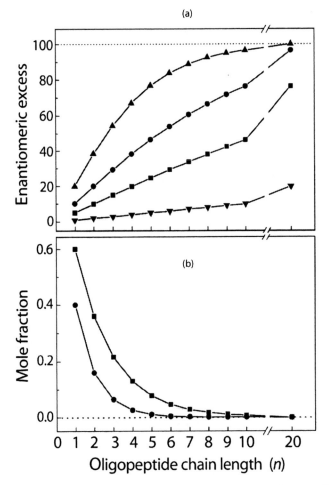

Figure 4.10 (a) Theoretical enantiomeric excesses, $[(L_n - D_n)/(L_n + D_n)] \times 100$, in the enantiomeric oligomers with homochiral sequence as a function of n, on starting the polymerization with various enantiomeric excesses, $[(L - D)/(L + D)] \times 100$, of the L-enantiomer: (▼), 1%; (■), 5%; (●), 10%; (▲), 20%. (b) Theoretical mole fractions of the enantiomeric leucine oligomers with homochiral sequence as a function of n, on starting the polymerization with an enantiomeric excesses of 20% of the L-enantiomer: (■), $(L\text{-Leu})_n$; (●): $(D\text{-Leu})_n$. (From Hitz and Luisi, 2004.)

is a trivial amplification of homochirality with increase in chain length simply on account of statistical laws (Bonner, 1999; Hitz and Luisi, 2004).

The relation between the single-chain homochiral amplification and the dilution is given in Figure 4.10 (Hitz and Luisi, 2004). Of course the L_n oligomers would become extremely diluted in the total mixture, so that from a practical point of view this 100% e. e. is not very useful. One would need some specific methods to extract these few macromolecules from the chaotic random mixture, and this is not easy. However these simple calculations show that homochiral long sequences can in principle be obtained starting from a minimal enantiomeric excess and by simply applying statistical laws.

Those are theoretical considerations. Let us take some experimental data on the relation between chain growth and chirality. One such reported effect is the case in which the initial growing oligomer is homochiral, and fed with a racemate. The chain might be expected to grow in an atactic manner, with L and D units inserting in a random way in the growing chain, as we have figured out in the previous ideal case. However, instead – this is the interesting observation – the chain tends to grow such that it maintains the same initial chirality. In other words, there is selection, as if somehow the chains like to maintain the same homochirality.

This has been observed by Eschenmoser's group in the case of oligonucleotides (Bolli *et al.*, 1997a, b). They used pyranosyl-RNA tetramers with sequences chosen to be partially self-complementary, such as ATCG. They showed that these oligomers can self-assemble in dilute solutions and that the ligation reaction is highly chiroselective, being slower by at least two orders of magnitude when one of the (D)-ribopyranosyl units of a homochiral (D)-tetramer-2′,3′-cyclophosphate is replaced by a corresponding L unit. The authors conclude (Bolli *et al.*, 1997a):

> Such a capability of an oligonucleotide system deserves special attention in the context of the problem of the origin of biomolecular homochirality: breaking molecular mirror symmetry by de-racemization is an intrinsic property of such a system whenever the constitutional complexity of the products of co-oligomerization exceeds a critical level.

The observation that chains tend to grow in a homochiral way, is also true for polypeptides. For example in the polymerization of NCA-activated α-amino acids there is a natural tendency to form homochiral chains (Blocher *et al.*, 2001; Hitz and Luisi, 2004). In particular when a racemic mixture of NCA-activated amino acids is polymerized, a mixture of chain lengths up to heptamers or decamers is obtained, which, as expected, is a racemate. However, in these products there is a significant excess of homochiral stretches. One typical result is shown in Figure 4.11.

One can see that the experimentally observed frequency of homochiral chains is significantly larger than that expected on the basis of an ideal random copolymerization of D and L enantiomers. The data shown in Figure 4.11 are relative to the case in which polycondensation is carried out in the presence of liposomes, but similar results are obtained also without liposomes. It is just that with liposomes longer oligomers are obtained, as the longer oligomers adsorb on the surface of the liposomes, and do not precipitate. Aside from this, the data suggest a kind of spontaneous homochiral self-organization. I find this important because it may

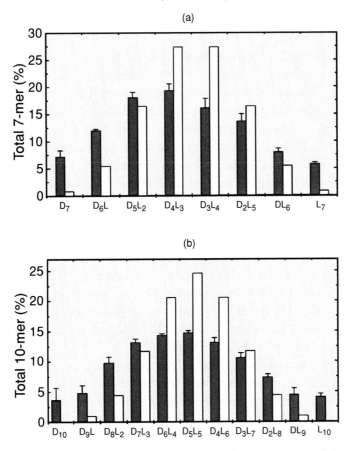

Figure 4.11   Relative abundance for D_pL_q-stereoisomer groups of the oligo-tryptophan n-mers ($n = 7$ and 10, respectively), obtained after two racemic NCA-Trp feedings (a) in the absence ($n = 7$) and (b) in the presence of POPC liposomes ($n = 10$). The relative abundances of the D_pL_q stereoisomer subgroups (dark-gray columns) are mean values of three measurements. Standard deviations are given as error bars. The white columns correspond to the theoretical distribution, assuming a statistical oligomerization. (From Blocher *et al.*, 2001.)

suggest a general principle of self-organization, according to which a small, initial structural order tends to be maintained and amplified.

The natural tendency of polypeptide chains to grow homochirally may suggest an alternative mechanism for the breaking of symmetry, based on macro-molecules instead of monomers. The argument is that it should be easier to separate enantiomeric homochiral chains, rather than racemic low-molecular-weight monomers, from each other. It has been shown for example that when the NCA-polycondensation is performed on mineral support, the oligomeric prod-uct remained absorbed on the surface. The lower oligomers are, however, easily eliminated on washing with water, whereas the higher oligomers remain absorbed (Hitz and Luisi, 2004).

Whereas this relation between chain length and homochirality is almost trivial, the following consideration, suggested by Eschenmoser (Bolli *et al.*, 1997a, b), is not. Consider a racemic monomer mixture, and assume that we are dealing with an ideal enantioselective oligomerization – one that produces mostly homochiral L- or D-sequences, as we have seen are possible in the examples above. The point is that over a critical chain length the product can no longer be a racemate. Above a certain length the constitution of the homochiral D-oligomers and L-oligomers cannot be identical. This is because the number of possible homochiral sequences with respect to the number of actually formed chains is so large that it is extremely unlikely for both a given D or L-sequence with length *n* to form, particularly when we have a limited supply of monomers. Also, if the constitution of D- and L-sequences is not the same, then there cannot be a racemate. This would be a kind of symmetry breaking brought about by the length of the chain, and which is also a consequence of the statistics of chain length. Of course, the question in a real case is whether the oligomerization suggested by Eschenmoser produces an excess of one type of sequence significant enough to be physically separated; the problem may be the extreme dilution indicated in Figure 4.11b.

There is another fascinating item in this general field of chirality of macromolecules, this time nothing to do with covalent chains, but to do with aggregate formation. This is symmetry breaking brought about by self assembly (Ribo *et al.*, 2001; Crusats *et al.*, 2003; Purrello, 2003; de Napoli *et al.*, 2004). This forms the subject of Chapter 5, concerned with self-organization.

Concluding remarks

Every one of us working in the field has some bias or other. One of mine concerns the question of macromolecular sequences. The bias is, that the bottom-up approach to the origin of life will never be close to a solution – both conceptually and experimentally – until the problem of the onset of macromolecular sequences is clarified. Obviously the origin of the specific macromolecular sequences (as opposed to simple polymerization) is not an easy question to answer, as it is linked to the general problem of structure regulation.

We have learned previously that the synthesis of exactly "our" proteins on Earth is doomed by contingency – we cannot hope to find out the exact conditions that determined the final sequence of a given protein or a nucleic acid from our Earth. Once this bitter assertion is accepted, we should at least attempt experiments that show that the prebiotic synthesis of *some* specific sequence in many identical copies is possible. Also, it is perhaps surprising that there has been so little reported in the literature about this: as already mentioned, co-oligopetides or co-oligonucleotides of say, 30 residues long, which have been produced under "honest" prebiotic condition have not yet been characterized. The same is also true, of course, for the "prebiotic" RNA world, which in my opinion is still mostly based on the naïve expectation that a self-replicating RNA family has "popped out" from the prebiotic soup. Research on this goes on and perhaps the molecular biologist's dream will come true. In the meantime we need some alternative working hypothesis, and this is the reason why I have presented a model for the aetiology of macromolecular sequences – something that can be proved or disproved experimentally.

The answer to the question of the aetiology of macromolecular sequences is elusive; as is the emergence of homochirality in the biopolymer chains. Have chains emerged when on Earth there was already an established homochirality of the biomonomers, or was the formation of long chains the cause of the emergence of homochirality?

Again, we do not know the answer, and so it is for most of the questions asked in the last two chapters under the logos "why this . . . and not that?" Despite the partial fog that we have received as an answer, I still believe that by asking such questions we are doing a good job – at least in clarifying the really important questions.

We are now ready to further escalate the ladder of complexity and investigate the notion of self-organization, emergence, and self-replication.

Questions for the reader

1. Would you encourage experiments according to the scheme developed in this chapter for explaining the emergence of macromolecular sequences? If not, what would you add/modify?
2. Are you in favor of a double origin of macromolecular sequences (one for proteins, one for nucleic acids); or do you believe that one derives from the other in a causal way (genetic code or something similar in primordial time)?
3. Do you see any way to make a stereoregular polypropylene displaying optical activity?

5

Self-organization

Introduction

In Chapter 1 we mentioned Oparin's bold idea that the transition to life was based upon a gradual and spontaneous increase of molecular complexity. This ordering process in a prebiotic scenario must have taken place without the intelligence of enzymes and without the memory of nucleic acids, as by definition these did not yet exist. At first sight, this whole idea appears then to be at odds with the second law of thermodynamics and the common belief that natural processes preferentially bring about an increase of entropy/disorder.

As a matter of fact, there are quite a few processes that bring about an increase of molecular complexity, the general term for such processes being self-organization. Some of these processes are under thermodynamic control, i.e., occurring with a negative free energy change; and there are also self-organization processes that are not spontaneous, being under kinetic control.

Together with this increase of structural complexity, another property must be considered – the notion of emergence. Although self-organization and emergence go hand in hand, for heuristic reasons emergence will be discussed in a separate chapter. A particular combination of self-organization and emergence gives rise to self-reproduction, and this will be discussed afterwards.

The terms self-assembly and self-organization are often used synonymously. One might argue that the term self-assembly is more general, and self-organization focuses on those cases of self-assembly that give rise to a significant degree or order (organization). In this text the term self-organization will mostly be used.

There is a vast amount of literature on self-assembly and/or self-organization, as this notion is used in practically all fields of science, from classic organic chemistry to polymer chemistry (Lindsey, 1991; Lawrence *et al.*, 1995; Pope and Muller, 1991; Zeng and Zimmermann, 1997), to the new frontiers of nano-technology, nano-robotics (Whitesides *et al.*, 1991; Bissel *et al.*, 1994; Whitesides and Boncheva,

2002), surface chemistry, and electrochemistry (Miller *et al.*, 1991; Dubois and Nuzzo, 1992; Ulman, 1996; Decher, 1997). Of course, practically all biology is characterized by self-assembly and self-organization, from protein folding to protein–protein interactions, from the duplex of nucleic acids to protein–nucleic acid complexes, from the assembly of actin and the formation of microtubules to the assembly of tobacco mosaic virus, as well as the assembly of a cell and organs. Among the many books devoted to this field, see, for example, those by Birdi (1999), Westhof and Hardy (2004), Riste and Sherrington (1996); and also a recent editorial by Glotzer (2004). It is not the aim of this chapter to make a general review of the field, but to give some general information and considerations that are relevant for the origin of life. In this context, several specific references will be discussed.

I believe it is useful to start with a qualification on the term "self". *Self*, in the connotation of this chapter and in the field of life science in general, defines a process that is dictated by the "internal rules" of the system. The term can be applied to the case of kinetic control as well, provided that the agents responsible for the kinetic control are considered as part of the system's "internal rules." Conversely, when the structure is organized by external, imposing forces, this is not self-organization. All this is rather trivial, however in the literature there are several cases where the term "self-assembly" is misused (for example, where the assembly is actually man-made, as in the example of layered nano-structures organized by a series of external manual operations). In Table 5.1 various examples are shown, and on the basis of the above considerations it is rather easy to distinguish which are cases of self-organization and which are not.

The simplest way in which a process occurs "by itself" is when it is under thermodynamic control. The folding of a protein, or the self-assembly of micelles at the critical micelle concentration (cmc) are examples of spontaneous processes; the latter are characterized by a negative free-energy change, as the self-organized product has a lower energy than the single components.[1]

In the examples given in Table 5.1, the assembly of a TV set, or the page numbering of a manuscript, or the growth of a city, are certainly not self-organization processes, as they are imposed from the outside. Protein folding, subunit interactions in oligomeric proteins, DNA duplex formation, micellization, are all examples of self-organization under thermodynamic control; whereas the examples of more complex biological systems (e.g., the assembly of a virus, of a bee hive, and

---

[1] Notice that a strict thermodynamic definition of a "spontaneous process" (for example see Atkins, 2002) refers only to thermodynamics – simply an exoergic process – without taking into account the kinetics. According to such a strict definition, a given process is "spontaneous" even when it does not occur, because of the high activation-energy barriers. For example, the formation of water from hydrogen and oxygen in the absence of catalyst is one such process. This definition is formally correct, but confusing from the practical point of view and in this text; as already emphasized elsewhere (Luisi, 2001) the term spontaneous characterizes a process that is not only thermodynamically favored, but that also occurs rapidly in the timescale of the observation.

Table 5.1. *Examples of self-organization; and not.*

Assembly of a TV	Queuing by the post office	Assembly of a virus
Swarm intelligence	Building of an ant hill	Assembly of ribosomes
City growth	Numbering of book pages	Formation of micelles
Protein folding	DNA duplex formation	Assembly of the cell
Crystallization	Oligomeric assembly of hemoglobin	Surface metal coating

all other social-insect constructions), being the result of genomic and enzymatic activity, should be considered self-organization processes, which are under kinetic control. The application of thermodynamic and kinetic concepts becomes difficult or impossible in the macroscopic and social domains; however, the criterion of whether the organization is the fruit of the internal rules of the system is generally always applicable. For example, the organization of a political party, or of a company, or the patterns of the swarm intelligence, can be considered as the outcome of the system's internal rules.

In the following section some of these examples will be discussed more specifically, and at the end of this chapter a classification of the main self-organization phenomena will be proposed.

Self-organization of simpler molecular systems

Micelle formation is a nice example of self-organization under thermodynamic control. Following the addition of some liquid soap in water at a concentration higher than the cmc, spherical micellar aggregates spontaneously form. This process takes place with a negative free-energy change – actually the process is attended by an increase of entropy.

This is based on the fact that amphiphilic molecules tend to aggregate in order to decrease unfavourable contacts with water molecules – thus, upon aggregation, water molecules are set "free" and this brings about an increase of entropy. This is qualitatively illustrated in Figure 5.1. The example illustrates beautifully, in its simplicity, that formation of *local order* and *increase* of the overall entropy can take place simultaneously. The organized, ordered structures are thus a kind of by-product of the overall increase of entropy ("disorder").

Of course, self-assembly of this kind occurs in water, and not, say, in ethanol. Any self-organization process must be defined in a given set of initial conditions. Initial conditions, as always in thermodynamics, determine the outcome of the process, and in particular whether the process is also under thermodynamic control or not. Aside from that, it is well known that a large series of amphiphilic molecules

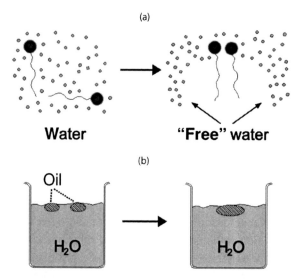

Figure 5.1 (a) The aggregation of surfactant molecules showing the increase of entropy due to the "liberation" of water molecules. Hydrophobic forces are the main factors for the association of surfactant molecules. (b) The analogy with two droplets of oil in water.

tend to aggregate, as shown in Figure 5.2. The later sections of Chapter 11 consider vesicles and liposomes; here it is important to anticipate that the formation of such aggregates can be attended by an additional type of ordering process: compartmentation and segregation of guest molecules. This is qualitatively illustrated in Figure 5.3: during formation of vesicles, depending upon their chemical nature, molecules can be sequestered on the bilayer, or in the aqueous interior, or on the polar bilayer surface.

In addition to surfactants, a reasonable number of synthetic and natural compounds tends to self-organize into bulk phases exhibiting periodic behavior as in crystals. As Zeng and others explain (Zeng *et al.*, 2004), the resulting periodical self-organization can be one-dimensional (lamellar phase), two-dimensional (columnar phase), or three-dimensional (micellar or bi-continous phase). The authors also investigate the assembly of dendrons and dendrimers (tree-like molecules); the corresponding nano-structures are shown in Figure 5.4.

Crystallization is another typical example of self-organization. The form of the crystal lattice, say of crystallizing NaCl, is not imposed by external forces, but is the result of the internal structural parameters of the NaCl system under the given conditions (temperature, etc.). Protein folding is also a self-organization process, determined by the internal rules of the system (primarily the primary structure). As is well known, Chris Anfinsen and coworkers demonstrated in the 1970s that

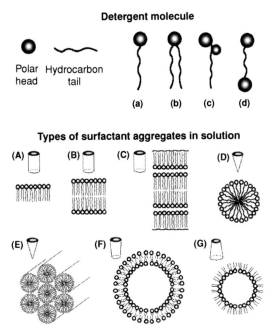

Figure 5.2 Top-diagramatic representation of a detergent molecule. (a) Single tailed; (b) double tailed; (c) zwitterionic; (d) bolamphiphilic. Bottom – different types of surfactant aggregates in solution: (A) monolayer; (B) bilayer; (C) liquid-crystallin phase lamellar; (D) normal micelles; (E) cylindrical micelles (hexagonal); (F) vesicles (liposomes); (G) reversed micelles.

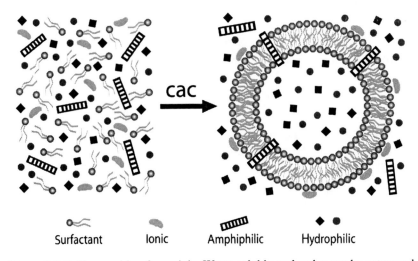

Figure 5.3 Self-assembly of a vesicle. Water-soluble molecules can be entrapped inside, ionic molecules on the polar head groups of the surface, amphiphatic molecules in the hydrophobic bilayer. (cac: critical aggregate concentration).

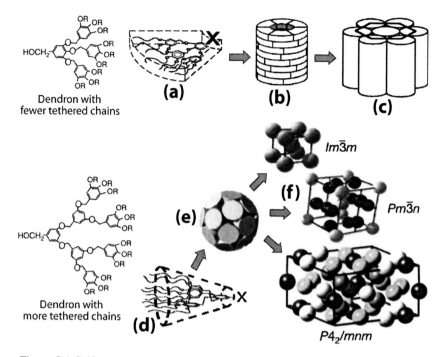

Figure 5.4 Self-assembly of wedge-shaped molecules (modified from Zeng *et al.*, 2004). (a) Dendrons with fewer tethered chains adopt a flat slice-like shape (X is a weakly binding group). (b) The slices stack up and form cylindrical columns, which assemble on a two-dimensional hexagonal columnar (Col$_n$) lattice (c). (d) Dendrons with more end-chains assume a conical shape. (e) The cones assemble into spheres, which pack on three different three-dimensional lattices (f) with different symmetries.

protein folding is a process under thermodynamic control. The criterion resides in the reversibility of folding in vitro: the protein is put in a glass vial, far removed form the magic of the cell, denatured with urea or whatever mild reagent, the denaturing agent removed by dialysis – to see whether the original native conformation is regained. If it is, the process of folding is under thermodynamic control, since only thermodynamic factors can be present in the in vitro system; see Figure 5.5.

This criterion is good for establishish whether a process is under thermodynamic control. Care should be taken however to understand the term "reversibility" in this case. The folding of a protein is generally per se a "chemically irreversible" process, in the sense that the chemical equilibrium is overwhelmingly shifted towards the folded form – there is not a low activation energy barrier between the native folded and the unfolded form and a corresponding chemical equilibrium in the native state between the two forms. Thus, in the case of the "thermodynamic hypothesis" of

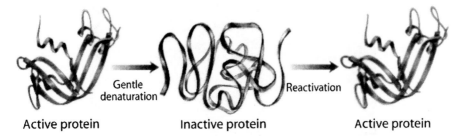

Active protein   Inactive protein   Active protein

Figure 5.5 Denaturation and reactivation. When a protein is denatured, it loses its normal shape and activity. If denaturation is gentle and if the conditions are removed, some proteins regain their normal shape. This shows that the normal conformation of the molecule is due to the various interactions among a set sequence of amino acids. Each type of protein has a particular sequence of amino acids. (Adapted from Mader, 1996.)

Anfinsen, the term reversibility does not refer to the fact that we have a chemical equilibrium between the folded and unfolded state at room temperature, but that after disruption of the initial state by a particular reaction, the original native folding by a spontaneous process can be reconstituted. Likewise, in the formation of liposomes from phosphatidyl surfactants, where the equilibrium constant is of the order of 10^{-10} M in favour of aggregation, there is no chemical equilibrium between disordered monomers in solution and liposomes – the activation energy to go back from liposomes to monomers is too high. However, we can also look at the liposome formation as a reversible process, in the sense that one can disrupt the aggregate with, say, addition of cholate, and than regain the original liposomes by removal of the added denaturing compound.

There are many other spontaneous associations of biopolymers in nature – such as the assembly of hemoglobin from the two $\alpha\beta$-dimers, and the more complex protein assembly process illustrated in Figure 5.6.

Not only proteins and their oligomers, but nucleic acids as well give beautiful examples of self-organization – think of the formation of the DNA duplex, where the primary structure of the two strands determines the rules for self-assembly; or the folding of t-RNA.

Self-organization and autocatalysis

Let's go back to much simpler systems, for example the self-aggregation of surfactant molecules. When surfactant molecules solubilize in water, often the process is slow at the very beginning, and gets faster with time: the more surface bilayer is formed, the more the process speeds up, because there is more and more active surface where the next steps of aggregation can take place. The same

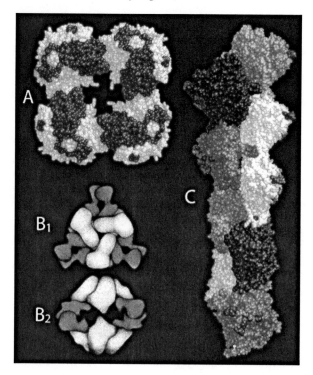

Figure 5.6 Self-organization in oligomeric proteins. (A) The transacetylase core of the pyruvate dehydrogenase complex. The core consists of 24 identical chains (12 can be seen in this view). (B) The aspartate transcarbamoylase, formed by six catalytic (lighter subunits) and six regulatory chains (darker subunits): (B_1) view showing the threefold symmetry; (B_2) a perpendicular view. (C) The helical assembly of several identical globular subunits in F-actin polymer. The helix repeats after 13 subunits. (All adapted from Stryer, 1975.)

behavior is observed in crystallization; in other words, an *autocatalytic process*: the product of the reaction (organized surface bilayer or crystals) speeds up further self-organization. Actually, the point can be made, that generally self-organization in chemical systems is attended by some kind of autocatalytic behavior. Some classic examples of self-organization are DNA self-assembly (see Figure 5.7), as well as the t-RNA folding (see Figure 5.8). This is so even for folding processes in proteins: once the first intermolecular interactions have been established, the next ones are more easily realized within the structure, as several chain segments are now closer to each other.

The relation between self-organization and autocatalysis is discussed in some detail by Burmeister (Burmeister, 1998), and that between chirality and self-organization/self-replication in biopolymers is considered from the theoretical point of view by Avetisov and Goldanski (1991).

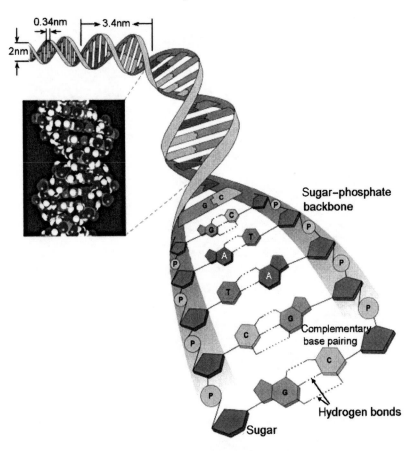

Figure 5.7 The well known principle of self-assembly of the DNA duplex.

Polymerization

Polymerization appears to be a simple method to increase molecular order: starting from a gaseous or liquid monomer mixture, long covalent macromolecules with a vast series of emergent properties can be obtained. At first sight, there is a strong similarity between polymerization and two other processes mentioned earlier, crystallization and surfactant aggregate formation. In all these cases, in fact, starting from a disordered mixture, the low-molecular-weight components are stringed together in a compact ensemble – a crystal, a micelle, a polymer chain. As shown in Figure 5.9, there is however an interesting difference between a polymerization process and a self-assembly process of surfactant aggregates: whereas the surfactant self-assembly is attended by an overall increase of entropy (because of water being made "free"), polymerization is generally attended by a decrease of entropy – as free monomers are being attached to each other in a covalent string.

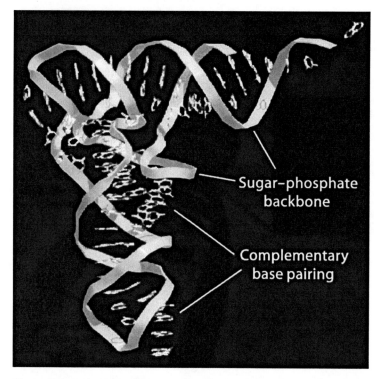

Figure 5.8 Folding of t-RNA, another classic case of self-organization.

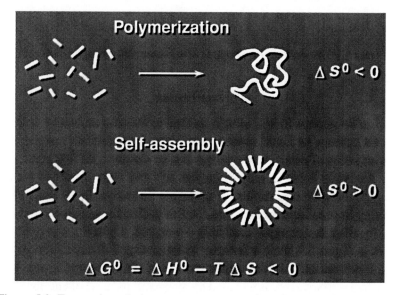

Figure 5.9 Two main ordering processes: comparison between the process of polymerization and the process of surfactant aggregation.

Is polymerization a self-organization process? Namely, is the product (the polymer) the result of the internal rules of the system? The answer is not as clear as in the case of the formation of a micelle or of a crystal. I believe that a positive answer can be given in the case of spontaneous polymerization, as in the formation of nylon starting from a mixture of dicarboxylic acid chloride and alkyl diamines; or the polymerization of styrene induced by heat or by light; or the formation of oligopeptides starting from *N*-carboxyanhydrides. However, in the case of step-wise polymerization – for example, in the Merrifield synthesis – this is not the case, and I would not include this polymerization process in the category of self-organization.

The simplest case of polymerization is the formation of a linear chain. There are several other stages of complexity in polymer chemistry, which include branching, cross-linking, polymer networks and dendrimers (for reviews see Hawker and Frechet, 1990; Föster and Plantenberg, 2002; Bucknall and Anderson, 2003; Pyun *et al.*, 2003). Not all these stages of complexity qualify as self-organization processes, as some are imposed by external forces and are not caused by the internal rules of the system. Perhaps the most interesting cases of self-organization are those obtained with aggregations of di-block polymers, as indicated by Bucknall and Anderson (2003). A di-block polymer has a chain consisting of two chemically different long sequences, of the type: ... AAAAAAAAAA–BBBBBBBBBBBB ... As in the case of surfactants, consisting of a polar and an apolar moiety, such block-polymers tend to phase-separate due to the immiscibility of the two blocks. A variety of structures can then be obtained by regulating the chemical nature and the length of the two ... A ... and ... B ... blocks. Some examples are shown in Figure 5.10. Block-copolymer self-organization occurs in dilute solution and also in pure bulk materials, and has already found technological application, for example for producing nanometer-scale templates, in solar cells and light-emitting diodes, and as photonic crystals (see specific literature in Bucknall and Anderson, 2003; Ma and Remsen, 2002). There is in fact intense and exciting research activity in this field, also due to the fact that these structures can be further stabilized after their self-assembly via covalent cross-linking – see for example the work of Ma and Vriezema, cited in Bucknall and Anderson, 2003.

Self-organization and kinetic control

Thus far, various cases of self-organization under thermodynamic control have been considered. There are also several cases in which self-organization is the result of kinetic control – in which the reaction product is not necessarily the most stable, but forms because of kinetic constraints. A simple qualitative illustration is given in Figure 5.11. The most general case of kinetic control is given by enzymatic reactions. In fact, Figure 5.11 is also qualitatively valid for a large number of

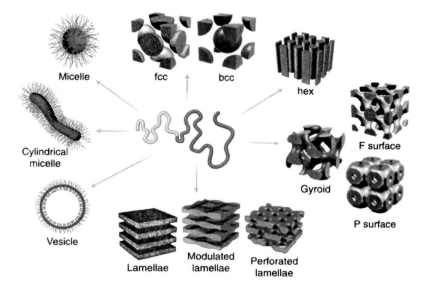

Figure 5.10 Self-organization of di-block copolymers. Block copolymers can form spherical and cylindrical micelles, vesicles, spheres with face-centered cubic (fcc) and body-centered cubic (bcc) packing, hexagonally packed cylinders (hex); minimal surfaces (gyroid, F surface, and P surface), simple lamellae and modulated and perforated lamellae. (Adapted from Bucknall and Anderson, 2003.)

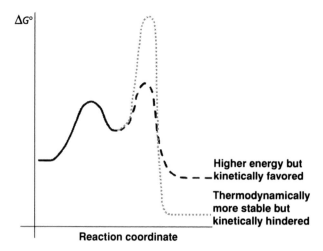

Figure 5.11 Qualitative illustration of the kinetic control in chemical reactions. Ziegler-Natta stereospecific polymerizations are examples of kinetic control. Most enzymatic reactions are kinetically controlled – those for instance which lead to stereoregular biopolymers.

enzymatic processes: once the intermediate (complex enzyme–substrate) has been formed, there is only one way to go, that with the lowest activation energy. Generally in fact the function of enzymes is to permit escape from the rigors of thermodynamic control, which would demand progression towards thermodynamic stability – or there is no reaction. Even for protein folding there are cases of kinetic control. Let's take the example of insulin. The folding of insulin is not under thermodynamic control, actually the biologically active folded insulin is a kinetically trapped form (Stryer, 1975). In fact, insulin is the product of the enzymatic cleavage of pro-insulin (and this in turn comes from pre-pro-insulin). In other words, the folding of insulin is not a spontaneous process but the result of enzymatic action, and is then under kinetic control.

Among the large number of reactions under kinetic control, let's consider those that lead to stereoregular biopolymers. As mentioned in the previous chapter, polysaccharides, polypeptides, and nucleic acids display several degrees of order and regularity. This extreme constitutional regularity and stereoregularity is due to a family of specific enzymes. Clearly the linear, stereoregular chains are not thermodynamically the most stable products; a random mixture of all possible enchainments would be entropically more favored, as it would correspond to billions of possibilities. Thus, the constitutional order and the stereoregularity of biopolymers results from a clear case of kinetic control. The same holds for stereoregular synthetic polymers, such as isotactic or syndiotactic polypropylene (refer back to Figure 4.9).

Also on the basis of this example, we can go back to the question of whether stereoregularity is a self-organization process. Again, the answer is positive, if the catalyst is considered as one of the determinants of the "internal rules" of the system. For example, if the growing polymer is considered as a complex with the catalyst (enzyme or Ziegler-Natta catalyst), then this can be deemed a self-organizing system. However, as in the general case of polymerization mentioned above, this is clearly a grey zone, where the notion of self-organization becomes less distinct.

Self-organization and breaking of symmetry

Self-assembly of chiral molecules may result in organized aggregates displaying a remarkable enhancement of optical activity. The best known examples are amino-acid residues that assume a periodic conformation – an α-helix or a β-sheet chain. In this case, the enhancement of optical activity is due to the onset of a particular rigid conformation.

There is a less trivial, actually quite fascinating area of inquiry in this general field of chirality of high molecular weight compounds; this is symmetry-breaking brought about by self assembly. One case in point is the aggregation of porphyrin-like compounds, as investigated independently by the research groups of Ribo

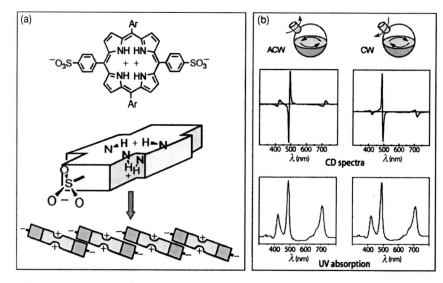

Figure 5.12 Chiral self-aggregation of porphyrin compounds (adapted from Ribo *et al.*, 2001, with kind permission). (a) The monomer and J-aggregates structures. (b), The outcome of rotating directions of the flask in the rotary evaporator, clockwise (CW) and anticlockwise (ACW), on the preparation of aggregates by concentration of a monomeric solution of the porphyrin. The corresponding CD spectra, showing the chirality signature, and the UV absorption bands of the J-aggregates are also shown. Notice that the two UV spectra are identical, whereas the two CD spectra are opposite to each other.

(Ribo *et al.*, 2001; Crusats *et al.*, 2003) and Purrello (Purrello, 2003; de Napoli *et al.*, 2004). Following the work of Ribo and coworkers, the achiral amphyphylic porphyrin building blocks (of the type shown in Figure 5.12) are observed to aggregate side-to-side in aggregates called J-aggregates. These aggregates form colloidal-like particles in aqueous solution and their shape depends on the substitution pattern of the porphyrin monomer. In the case of phenyl-and tris(sulfonatophenyl)-substituted porphyrins, the chiral polarization exerted by the direction of the stirring vortex in the rotary evaporator, during the concentration of the solution in the preparation of the aggregates, is strong enough to select the chiral sign of the aggregates. This should be attributed to the different growth of homochiral helices, which would show non-enantiomeric trajectories in a chiral hydrodynamical flow. As Ribo says (personal communication, 2004):

The effect probably can only occur when the process of growth of the mesophases is a cluster to cluster process and not a molecule to aggregate process. This work suggests that in supramolecular systems chiral long range forces can transfer information to the bond-length scale of size. This suggests that the role of vortices should be taken into account as one of the chiral polarization forces directing the chirality of spontaneous symmetry breaking.

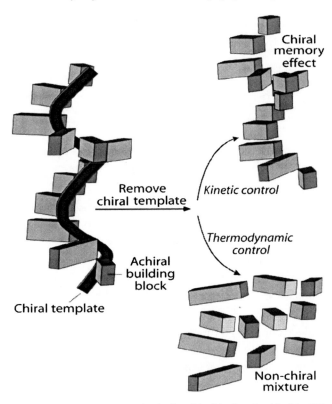

Figure 5.13 A supramolecular species built with chiral and achiral building blocks can retain the memory of the imprinted chirality (handedness) even after substitution or removal of the chiral component. (Adapted from Purrello, 2003.)

Ribo also suggests that this might be relevant in a prebiotic scenario, at the early stages of the origin of life, since there may well have been vortices determining permanent sign directions in primordial times.

In the same area, Purrello and Lauceri are also working with porphyrins. Similarly, in this case the supramolecular chirality derives from the non-symmetric arrangement of molecular components in a non-covalent ensemble. In such cases, dissociation may lead to the formation of a mixture of left- and right-handed optical isomers, forming a non-optically active racemic mixture. In order to prevent the formation of racemates these authors have imprinted the desired chirality onto the supramolecular architecture by using a chiral template. If the final product is the most thermodynamically stable one, substitution or removal of the chiral template results in a loss of supramolecular chirality (see Figure 5.13, lower part). However, if the structure is only kinetically stable, then chirality will persist for hours, months or even longer – even in the absence of the chiral template. This phenomenon is referred to as a "memory" effect (see Figure 5.13, upper part). This

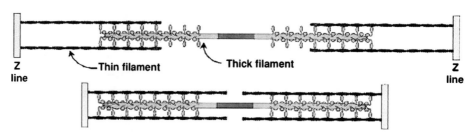

Figure 5.14 Schematic diagram showing interaction of thick and thin filaments in skeletal muscle contraction. (Adapted from Stryer, 1975.)

is a good illustration of the interplay between the two effects discussed, kinetic and the thermodynamic control in self-organization. In the case of Purrello, the species formed are not the most thermodynamically stable, but they are stabilized by kinetic effects (Purrello, 2003).

A beautiful example of this memory effect due to kinetic factors has been reported by Raymond and coworkers (Ziegler *et al.*, 2003). These authors describe the synthesis of kinetically stable chiral architectures built entirely from achiral structural elements. To induce chirality in this non-covalent assembly, they exploit the chiral arrangement of specifically designed achiral ligands surrounding metal ions. This complex is able to preserve its chiral memory despite the kinetic lability of the metal–ligand bonds, while at the same time allowing for dynamic ligand substitution.

The growing interest in supramolecular chirality stems not only from the intrinsic relevance of such studies for the origin of chirality in life processes, but also from the potential technological applications, such as the separation of optical isomers for the pharmaceutical or food industries.

Complex biological systems

The biological world is by definition full of self-organized structures, and here only a few examples are given. Usually, a complex interplay between thermodynamic and kinetic control is at work to guarantee the complexity of the biological structures. In addition, many such syntheses in vivo take place on a matrix – pre-existing fibers or membrane structures or organelles – so that steric factors also play a role in the assemblage. These steric factors can also be seen as determinants for the kinetic control. All this can be evidenced by the example of muscle-fiber organization (see Figure 5.14). Muscle fibers consists of thick and thin filament, and each type of filament is a conglomerate of different proteins (Stryer, 1975; Alberts *et al.*, 1989). Let us start with the thin filaments: the globular F-actin polymerizes to give a long double helix, a polymerization that can take place in the test tube and is therefore

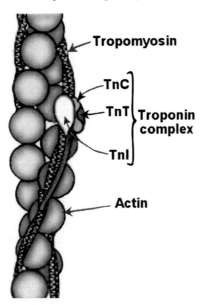

Figure 5.15 Model of a thin filament in muscle fibers: TnC, TnT, and TnI are all different forms of the protein troponin. (Adapted, with modifications, from Stryer, 1975.)

per se under thermodynamic control. Figure 5.15 shows the complex structure of a thin filament, and it can be observed that around the double helix of actin there is another double helical structure of a filamentous protein called tropomyosin. In the grooves of this complex helix, there is yet another globular protein, troponin (in turn formed by three different constituents).

Thus, the thin filament is really a quite complex interplay of self-organization structures locked one into the other. Turning now to the thick filaments, the main element is myosin, which comprises two heavy α-helical chains (each about 2000 amino-acid residues long) and four light chains, localized in the myosin head. This is illustrated in Figure 5.16. The two chains coil around each other to form a coiled coil, which is stabilized mostly by hydrophobic interactions between the two α-helical heavy chains. Hundreds of myosin molecules assemble together to build thick filaments mostly due to ionic interactions between the tails of the individual molecules (Alberts *et al.*, 1989). It is possible to induce spontaneous aggregation and disaggregation of myosin molecules by varying the salt concentration. The overall assembly process of thick and thin filaments to build the muscle fibers is certainly very complex, and it is difficult to discriminate the relative importance of thermodynamic and kinetic effects (Alberts *et al.*, 1989).

Thermodynamics has in this case to do with the specificity of the protein structure that determines the binding selectivity to one another, for example actin and

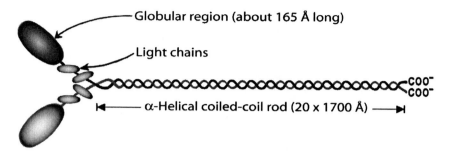

Figure 5.16 Organization of the thick filament in muscle-fibre proteins. (Adapted from Stryer, 1975).

tropomyosin. However the assemblage is also determined by the sequential steps of the enzymatic syntheses – and also by steric factors due to fiber matrices.

Another complex macromolecular aggregate that can reassemble from its components is the bacterial ribosome. These ribosomes are composed of 55 different proteins and by 3 different RNA molecules, and if the individual components are incubated under appropriate conditions in a test tube, they spontaneously form the original structure (Alberts *et al.*, 1989). It is also known that even certain viruses, e.g., tobacco mosaic virus, can reassemble from the components: this virus consists of a single RNA molecule contained in a protein coat composed by an array of identical protein subunits. Infective virus particles can self-assemble in a test tube from the purified components.

However, this cannot be generalized. For example the formation of some other bacterial viruses is determined by key enzymatic steps, and an example of this complex behavior is given in Figure 5.17, which illustrates the fascinating assembling process of the T4 phage. Here again there is a combination of kinetic and thermodynamic control: in fact, once the single components (the head, the tail, the tail fibers, etc.) have been made via kinetically controlled syntheses, most of the following assembly steps are determined by thermodynamic control. In this case, the mechanical mixing in a test tube of the single components would not form the original infectious virus. Also, certain cellular structures are not capable of spontaneous self-assembly, such as a mitochondrion, a cilium, or a myofibril, as part of the information for their assembly is provided by special enzymes. Again the term self-organization can also be used in these cases, whereby "self" indicates that the sequential and spatial order of this organizational structure is due to the internal rules of the system. In this case, the internal rules are determined by a large number of ordering factors, including enzymes – which, again, can be considered part of the system.

The same considerations hold for another beautiful object, depicted in Figure 5.18: an axoneme, from a bacterial flagellum (Stryer, 1975). This is really a

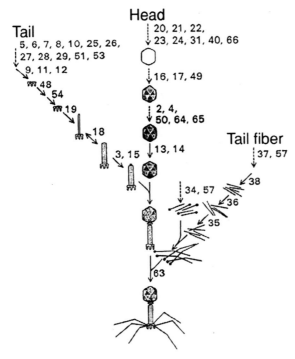

Figure 5.17 The assembly of a T4 phage from its constituents. The numbers next to the arrows refer to gene products that are required for a particular step in assembly. (Adapted from Wood, 1973.)

wonderful construction, a molecular motor that implies the collaboration of dozens of different specialized parts, each part being the organized structure of a bundle of specific proteins. Some may think of it as an oriental mandala. The perfect juxtaposition of all these components is the result of a calibrated, sequential ordering, synthetic, and organization process, where again thermodynamic and kinetic factors exert a concerted mechanism of utmost complexity.

A quite different example, that was considered in Table 5.1, is the formation of an anthill: the design of the anthill is not imposed externally; it is the result of the internal structure of the ant social system. It is self-organization, if we consider the ants themselves and their genome as part of the internal rules of the system.

Likewise, probably there is a very complex genomic determination in the case of "swarm intelligence:" with bird migration, for example, very regular patterns are formed (see Figure 5.19). It is known that the flying-formation pattern is not determined by the leading bird, as this can interchange. Thus, the regular pattern is a case of self-organization realized without any imposing external rules, similar to the anthill or the beehive.

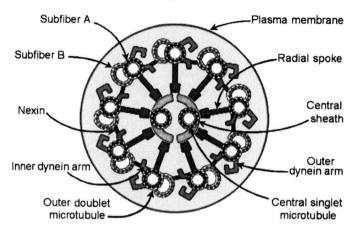

Figure 5.18 Schematic diagram of the structure of an axoneme – or a mandala? (Adapted from Stryer, 1975.)

Figure 5.19 Swarm intelligence in the case of migratory birds.

Again, we have a beautiful self-organization pattern, but no localized centre that determines the organization. The ordered complexity is an emergent, collective property, and one might even draw an analogy between the swarm intelligence and the formation of a micelle: the parts come together and form an ordered ensemble: in one case the self-assembly is determined by simple thermodynamic driving forces (hydrophobic interactions, etc.), in the other, complex genomic and social factors are at work. Despite the staggering difference in complexity, the underlying principle in both cases is the formation of a collective ensemble without an ordering centre or localized intelligence.

We will come back to this interesting point in the next chapter, which considers emergent properties.

Self-organization and finality

The pictures of the swarm intelligence, an axoneme or an anthill bring up an old question – the question of finality. It may in fact be argued that these complex biological systems, in contrast to micelle or to crystal formation, appear to have a rather specific finality. The relation between self-organization of the living and finality (also referred to as teleology, and/or teleonomy) is an old and complicated issue, which would involve a rather in-depth discussion on philosophy. The question of finality has been tackled for example by Monod (1971), who has introduced in this respect the notion of teleonomy to indicate an activity that is directed towards the realization of a biological program. The most salient program is the genetic one, which implies the species reproduction and evolution. Monod's term "program" is very important in this context, as it does not imply finality towards a future state, but rather a series of events coherent with and determined by the program itself. All subsystems, activities and structures that contribute to this main program are also teleonomic, in the sense that they are aspects and fragments of the main unique program. This idea is in keeping with the evolutionistic view of the complexity of formation; according to this the "watch" of the watchmaker is not a single act of one-instant creation by a constructor, but the result of a long series of consecutive small steps, each contributing gradually to the increase of complexity under the evolutionary pressure. On the other hand, some philosophers consider Monod's view an "escamotage" (Lazzara, 2001); a way to resolve with syllogisms the "hairy" question of finality.

In most classic biology literature, see for example Ernst Mayr (1988), the point is made that *teleology* is due to a distinct character of the organism, which is a program based on genetic information. The question of teleology is debated also within the theory of autopoiesis. It is argued that autopoiesis promotes the notion of an "internal teleology" (Weber, 2002), according to which finality is brought about by the internal logic of the system. This is conceptually not very different from the view expressed above by Monod.

This is all very interesting and is linked to the more general issues of philosophy of science. I will not dwell on these issues here, however I would like to make a remark that in my opinion is useful: that the notion of finality implies an observer – namely somebody who gives a valued judgment on the event. This valued judgment is thus an imposition, is generally context-dependent and may vary from observer to observer depending on the conceptual framework. Clearly, the ants do not know why they are making their anthill, and this construction may be viewed as an evolutionary

Figure 5.20 The artificial beehive structure obtained by heating a silicon oil between two glass plates (Bernal convection). (Adapted, with a few modifications, from Coveney and Highfield, 1990.)

optimisation device for preserving the species. On accepting this view, the question of finality may be laid to rest.

Out-of-equilibrium self-organization

The chemical and biological examples of self-organization given above are familiar to chemists and biologists. For a relatively large community of scientists working in the area of physics of dynamic systems, as well as artificial life, self-organization (and emergence) is, however, something else: something that is a characteristic of dynamical systems, namely systems that change over time. Terms such as chaos and non-linear dynamics, self-organized criticality, self-organization in non-equilibrium systems, are typical in this field (Nicolis and Prigogine, 1977; Bak *et al.*, 1988; Langton, 1990; Hilborn, 1994; Strogatz, 1994). Such dynamical systems are generally out of equilibrium, and at first sight it is counter-intuitive that a system out of equilibrium may form self-organized structures. This is in fact the challenge and beauty of this particular field.

One classic example is the formation of "beehive" structures in a liquid layer that is heated between two glass surfaces. Resulting convection (the so-called Bernal convection) brings about long-range hexagonal structures, instead of a more and more disordered molecular mixing, as to be expected. This regular, beautiful structure remains visible by the naked eye as long as the temperature difference is maintained; see Figure 5.20. The temperature at which the hexagonal cells form is a critical point, or bifurcation point, from which the system may adopt one pattern or another one (see Coveney and Highfield, 1990).

This notion of bifurcation point, connected with those of instability and fluctuations, is the basis of this branch of science of self-organization in out-of-equilibrium systems. The story begins with Alan Turing, who, in search of the chemical basis of morphogenesis, predicted that while homogeneity is the normal situation near equilibrium, systems may become unstable far from equilibrium because of fluctuations (Turing, 1952).

The next two important steps in this narrative are considered to be the following: (i) the description of the "bruxellator" by Prigogine and Lefever, who, following on from Turing's work, analyzed theoretically the ingredients that should be present in a model of chemical reactions in order to produce spatial self-organization (Prigogine and Lefever, 1968); (ii) the description of the Belousov–Zhabotinsky (B–Z) reaction.

The work of Prigogine and Lefever, in total observance with the second principle of thermodynamics, showed that regular oscillations of concentration had a sound thermodynamic basis, which were exhibited by the "bruxellator". In particular, the "bruxellator" shows how it is possible that order originates from the disorder throughout self-organization, which in turn is due to oscillations in a system out of equilibrium. Also, if the system is maintained far from equilibrium via a continuous addition of colored chemicals, the system can oscillate between two or more colored states. This is a particular case of dissipative structures giving rise to self-organization, the important concept developed by Prigogine and his school of thought. The notion of bifurcation also comes from this school, and the relevance to what is being discussed is illustrated in Figure 5.21. Here, the property λ (in the ordinate) at equilibrium, or close to it, obeys a linearity regime and the principle of minimal entropy production (another theorem of the Prigogine school) and the system is in a stable stationary state.

However at a critical distance from equilibrium, the system must "choose" between two possible pathways, represented by the bifurcation point λ_c. The continuation of the initial pathway, indicated by a broken line, indicates the region of instability. The concentration of the species A and the value of λ assume quite different values, and the more so, the further from equilibrium. An important point is that the choice between the two branching directions is casual, with 50:50 probability of either. The critical point λ_c has particular importance because beyond it, the system can assume an organized structure. Here the term self-organization is introduced as a consequence of the *dissipative structures*, dissipative in the sense that it results from an exchange of matter and energy between system and environment (we are considering open systems).

It is worthwhile also mentioning that the origin of homochirality has been viewed in terms of a bifurcation scenario (Kondepudi and Prigogine, 1981; Kondepudi *et al.*, 1985). In this case, the homochirality present on Earth would be a product of contingency.

These organized structures can take the form of *oscillations*, and this is indeed the case for the Zabotinski–Belousov (Z–B) reaction, observed in the 1950s by the Russian chemist Boris Belousov (see in Winfree, 1984). It is interesting to note

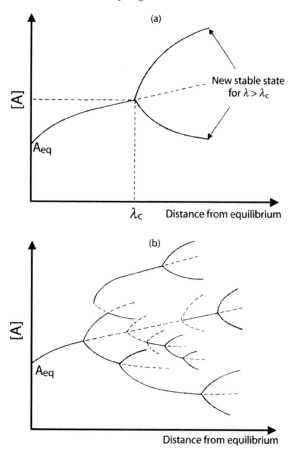

Figure 5.21 Bifurcation far from equilibrium. (a) Primary bifurcation: λ_c is the distance from equilibrium, at which the thermodynamic branching of minimal entropy production becomes unstable. The bifurcation point or critical point corresponds to the concentration λ_c. (b) Complete diagram of bifurcations. As the non-linear reaction moves away from equilibrium, the number of possible states increases enormously. (Adapted, with permission, from Coveney and Highfield, 1990).

that Belousov – just like Turing – was investigating a biological model, the Krebs cycle in his case. The complex reaction mixture contained potassium bromate (to simulate citric acid), sulfuric acid, cerium ions (to simulate the action of certain enzymes), and, as it was shown by later studies, may involve about thirty chemical intermediates and subreactions, some of them of autocatalytic nature (Winfree, 1984). Aside from the complexity of the reaction, which is not considered here, it is important to go back to the original observation of Belousov (later studied in detail by Anatoly Zhabotinsky): the solution began to oscillate, with great precision, between a colorless state and an orange one; and also the formation of particular

Figure 5.22 Some aspects of the Z–B reaction. (Adapted from Coveney and Highfield, 1990).

spatial structures was observed (see Figure 5.22). This was in keeping with the prediction by Alan Turing, who in the meantime had committed suicide. As remarked by Coveney and Highfield (1990), it is too bad that the two scientists never met.

The field of *oscillating reactions*, or periodic reactions, or chemical clocks, came out of this background; indeed quite a number of chemical systems have been described, which show this oscillating, periodic, regular behavior (Field, 1972; Briggs and Rauscher, 1973; Shakhashiri, 1985; Noyes, 1989; Pojman *et al.*, 1994; Jimenez-Prieto *et al.*, 1998).

How relevant is this phenomenology of out-of-equilibrium self-organization for life? For the scientists working in this field, the answer cannot be but positive, since all living systems are open, far-from-equilibrium, dynamic, non-linear, and dissipative structures. The complex adjectivation used in this last sentence is in fact typical for Prigogine's theory applied to living systems. To this list could be added irreversibly evolving systems, characterized by an "arrow of time," as Coveney and Highfield (1990) remark.

Another important message emerging from these studies is that order and self-organization can indeed originate by themselves – again suggesting that complexity can be a spontaneously occurring process.

We will come back to the questions of non-linearity and complexity when considering the notion of emergence, in the next chapter.

Concluding remarks

In order to summarize the various aspects of self-organization, the following classification can be proposed:

1. Self-organization systems under thermodynamic control (spontaneous processes with a negative free-energy change), such as supramolecular complexes, crystallization, surfactant aggregation, certain nano-structures, protein folding, protein assembly, DNA duplex.

2. Self-organization systems under kinetic control (biological systems with genomic, enzymatic and/or evolutionary control), such as protein biosynthesis, virus assembly, formation of beehive and anthill, swarm intelligence.
3. Out-of-equilibrium systems (non-linear, dynamic processes), such as the Zabotinski–Belousov reaction, and other oscillating reactions; bifurcation, and order out of chaos; convection phenomena; tornadoes, vortexes
4. Social systems: (human enterprises that form out of self-imposed rules), such as business companies, political parties, families, tribes etc.; armies, churches.

Point number four, about social systems, has not been considered here. This simple classification conveys well the complexity of the field of self-organization. The question of whether the human social organizational systems are genetically determined – sociobiology, as in the case of social insects – or induced by social and educational constraints is quite interesting, but again out of the limits of this chapter.

As already mentioned, self-organization processes give a kind of "free ticket" to move upwards in the ladder of complexity. However, is this really enough to reach the point of building macromolecular sequences and the first self-reproducing protocells?

This question is linked to that asked in the previous chapter, relative to the onset of products under kinetic control, such as ATP and, at another level, specific macromolecular sequences. In fact, the critical point in the origin-of-life scenario is the emergence of kinetic control in chemical reactions. Can this property emerge spontaneously from a scenario of reactions under thermodynamic control? Or is it too much to expect from emergence?

Emergence is the subject of next chapter – we will see what emergence can do; and what it cannot do.

Questions for the reader

1. Do you accept the idea that self-organization in prebiotic time was the main driving force for the formation of the first living cells? (If not, what would you add to the picture?)

2. Suppose you divide a prokaryotic cell into its components, say ten different fractions, obtained by mild procedures; and then mix them all together. Would the living cell self-organize again? If not, why not? Also, which kind of cell would you choose to run this kind of experiment?

3. Is the folding of proteins activated by chaperons under thermodynamic, or under kinetic, control?

4. Are you convinced of the fact that finality is not an issue in the field of self-organization?

6

The notion of emergence

Introduction

In the previous pages we have discussed how the increase of molecular complexity may proceed via self-organization, emphasizing, however, that a simple increase of size and/or complexity is not enough – this must be accompanied by the onset of novel properties – up to the point where self-reproduction and eventually life itself arise. In fact, self-organization must be considered in conjunction with corresponding emergent properties.

The term "emergence" describes the onset of novel properties that arise when a higher level of complexity is formed from components of lower complexity, where these properties are not present. This is often summarized in the popular assertion that the "whole is more than the sum of the parts", and/or with the vague term "holism".

For a long time, emergence has been an active field of inquiry in the philosophy of science. As noted by McLaughlin (1992), the work of "British emergentism" can be dated back to Mill (1872) and Bain (1870) and flourished in the 1920s with the work of Alexander (1920), Morgan (1923), and Broad (1925); and the inquiry continues up to the present (see, Wimsatt, 1972; 1976a, b), Kim, 1984; Klee, 1984; Sperry, 1986; O'Connor, 1994; Bedau, 1997; Farre and Oksala, 1998; Holland, 1998; Primas, 1998; Schroeder, 1998, and several others).

The possible relevance of chemistry in the notion of emergence was realized as early as the mid-nineteenth century (Mill, 1872); and in 1923, as quoted by McLaughlin (1992), Broad stated that:

the situation with which we are faced in chemistry (. . .) seems to offer the most plausible example of emergent behavior.

In our times, in addition to chemistry and biology, emergence has been discussed in quite a variety of research fields, such as cybernetics, the theory of

112

complexity, artificial intelligence, non-linear dynamics, information theory, music, and social systems. Due to such a disparity of origins, it is not surprising that emergence often has a confusing connotation. The term "supervenience" is also used.

The reason a full chapter is devoted to emergence, lies in the fact that the notion of emergence connects chemistry and biology to the philosophy of science, and yet this argument is usually dealt with only in the specialized philosophical literature, which young scientists usually do not read.

Let me start by saying that, just like many arguments of philosophy of science, the notion of emergence can be considered on two different levels. On the one hand there is the "ontic" interpretation, which refers to a theory about things as they really are, independent of any observational or descriptive context. On the other hand, there is the "epistemic" interpretation, which refers to our knowledge of observable patterns or modes of reactions of systems (Primas, 1998). Whereas the first level makes a direct application to practical problems rather difficult, an epistemic description allows better contact with empirical reality. This chapter deals with the second, and for me easier, epistemic approach, with the expectation that this would be more appealing and comprehensible to graduate students of science (see also Luisi, 2002a, b).

Another qualification concerns the term "complexity," which is often used in relation to emergence. Again, an attempt to define complexity would lead us astray, and an interested reader is referred to the specialized literature, for example Weaver (1948), Platt (1961), Wimsatt (1972, 1976a and b), and Baas (1994). For the purposes of this chapter, it will be sufficient to work within the relatively simple framework proposed by Simon (1969). Accordingly, a complex system is seen as a hierarchic system, i.e., a system composed of subsystems, which in turn have their own subsystems, and so on. Consider, for example, the hierarchic progression from subatomic particles to atoms, molecules, and molecular complexes; or the progression from cells to tissues, organs, and organisms; the cell itself being composed of subsystems such as mitochondria, microsomes, the nucleus, etc., each of these particles being in turn composed of smaller entities. Or, leaving the field of science, consider the progression from a room to an apartment to a house to a block to a city, or from an individual to a family to a clan to a city population to the population of a nation . . . Each of these hierarchic levels has its own autonomy with respect to the higher and lower levels.

Again, this is an epistemic approach and a more rigorous approach in the philosophy of science would consider this too simplistic, as it is based on the assumption that objects and levels of structures can be considered as *separated*.

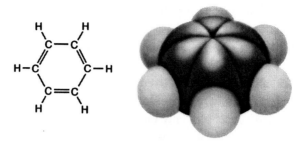

Figure 6.1 The aromatic character of benzene (deriving from the conjugation of double bonds) is an example of an emergent property.

A few simple examples

The trivial chemical example that is usually given in the literature on emergence since the early times of the British emergentists is water being formed from its atomic components. The collective properties of water are not present in hydrogen and oxygen; so the properties of water can be viewed as emergent ones.

This kind of simple argument can be generalized to encompass all cases in which a molecule is formed from its atomic components, e.g., CH_4, CO_2, HCl, NH_3, and so on. In each case, obviously, the properties of the resulting molecule are not present in the initial components and can be seen as emergent properties.

Francis Crick, in his book "The astonishing hypothesis" (1980), stresses the concept that there is nothing particularly new or exotic in the notion of emergence, as chemistry is full of it. He gives, as one of many, the example of benzene. The aromatic character of the benzene molecule is obviously not present in the atomic components, but is a property arising in the ensemble of the particular atomic configuration – an emergent property (see Figure 6.1).

Turning to a more complex chemical structure, such as myoglobin (Figure 6.2a), the specific binding properties towards the heme group are obviously not present in the single amino acids – the binding specificity is an emergent property of the particular molecular ensemble.

Going from myoglobin to hemoglobin (Figure 6.2b), a higher level of hierarchic structure is found, as haemoglobin is formed by four chains, each of them very similar to that of myoglobin. In this case a new quality arises from this assembly: whereas the binding of oxygen to myoglobin (single chain), gives a normal hyperbolic saturation curve (Figure 6.2c), in the case of mammal hemoglobin the binding isotherm is sigmoid. The difference is the very basis of respiration in mammals and is due to the cooperativity of the four chains in hemoglobin, which can be viewed as an emergent property arising from the interaction of the four chains.

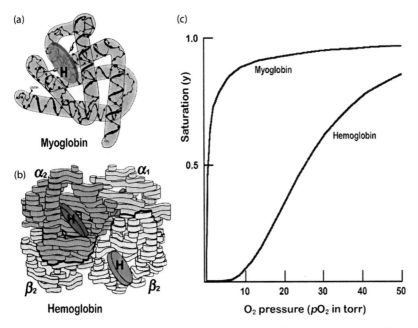

Figure 6.2 The single chain of myoglobin (a) and the four chains (two α and two β) of hemoglobin (b). The cooperative behavior of hemoglobin (sigmoid curve), as indicated in the left panel (c), can be seen as an emergent property, resulting from the interaction of the four chains.

The emergence of novel properties due to self-assembly is also present in much simpler systems. Consider, for example, the formation of micelles and vesicles from surfactants, as already seen (Figure 5.3).

A simple increase of molecular weight or size is obviously not an emergent property per se; however, the compartmentation, the constitution of an internal water pool which is distinct from the outside, is one such property. There are also collective motions of aggregates, as well as changes in the physical properties – such as the pK of fatty acids constituting an assembly – that are clearly emergent properties of the aggregate. Collective properties, which are characteristic of whole aggregate systems, have also been noted by Menger (1991).

As already mentioned in the introduction, the notion of emergence is applicable to many other fields. In the preface to the ECHO III conference, edited by Farre and Oksala (1998), applications to music, language, painting, memory, biological evolution, and the nervous system, are mentioned and discussed. Taking instead extremely simple examples, consider the geometric emergence shown in Figure 6.3: it is clear that the notion of angle has no meaning at the hierarchic level of the lines; likewise, there is no notion of surface at the hierarchic level of angles: and this flat world of surfaces does not have the property of volume. Thus,

Geometric emergence

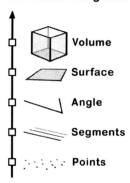

Figure 6.3 Emergence in simple geometrical forms. At each higher hierarchic level, novel properties appear that are not present in the components of the lower levels.

at each increasing step of complexity a novel feature, an emergent property that is not present in the lower hierarchic level, is found.

One question that can be posed after analysing these few examples is whether, and to what extent, the emergent properties can be interpreted in terms of the elements of the lower level, or, more generally, what is the relation between the emergent properties and the properties of the more basic level.

Emergence and reductionism

This question brings us directly to the relation between emergence and reductionism, which is another complex topic, abundantly discussed in the specialized literature, see Schroeder (1998), Wimsatt (1972), and Primas (1985, 1993, 1998).

Generally reductionism and emergence are presented as two opposite fronts: whereas emergence deals with the onset of novel properties, which are not present in the basic components, and as such has an upwards direction, reductionism generally looks down from a certain level of complexity claiming to explain each level on the basis of the lower ones. There are of course different positions within reductionism, including cases of radical reductionism. For example the more radical views of Oppenheim and Putnam (1958), Nagel (1961), and Reichenbach (1978), all cited by Atmanspacher and Bishop (2002), namely that chemistry and all natural science can be reduced to physics, are strongly opposed by Primas (1993). The strongest form of reductionism maintains (according to Ayala, 1983, and cited by Primas, 1998):

that organisms are ultimately made up of the same atoms that make up inorganic matter, and of nothing else.

Here it is important to make a clarification: the main point of opposition between emergence and reductionism concerns the problem of properties. There is nothing wrong, in principle, with reductionism when it stops at the level of structure: we can all agree that water consists of hydrogen and oxygen. The problem with reductionism is with the claim that the properties of water can be reduced to the properties of hydrogen and oxygen, and those to the properties of more elementary particles.

As pointed out by Wimsatt (1972), it is possible to be an emergentist and a reductionist at the same time, accepting the reductionistic view in terms of structure, and the emergentistic view with regard to properties. On the other hand we probably all agree that the level of structure alone is not very satisfactory. Returning to the previous assertion by Ayala that all living systems consist of the same atoms that make up the inorganic matter, we can even agree with him. However, what is completely missing in this view is the notion of emergence at the very many levels of increasing complexity – from atoms to molecules, from molecules to genes and enzymes, from these to cells, from cells to organs, whereby each time there is the onset of novel properties that are not present at the lower level of complexity. Thus, a rose can be said to be constituted by atoms and molecules, but the essence of a rose cannot be interpreted in terms of atoms and molecules – unless, to take the extreme reductionistic view, the properties of each level of complexity can be explained each time in terms of the properties of the lower level. This is however the very reductionistic notion that is very difficult to defend; and this brings us to the next point.

Deducibility and predictability

We can now focus on the specific question, can the emergent properties be deduced from the properties of the components?

The relation between emergent properties and the properties of the components has two sides. One can ask whether the properties of water (or any other molecule) can be explained a posteriori in terms of the properties of the components; or instead pose the question of whether the emergent properties can be foreseen a priori from the properties of the components. Suppose that we have only hydrogen and oxygen (we do not know anything about the existence of water). Can the formation of water and its properties be predicted? If we have only amino acids, can the existence of myoglobin and its properties be predicted?

The distinction between deducibility and predictability is important. Consider, for, example in biological evolution the emergence of flagella in bacteria, or wings in the early birds. The development of such properties cannot generally be predicted. However, once they are there, they can be deduced a posteriori from a series of small evolutionary hints.

This relation between emergent properties and properties of the basic components has been much debated in the literature. One school of thought claims that the properties of the higher hierarchic level are *in principle* not deducible from the components of the lower level. This is the so-called "strong emergence" or radical emergence, that demands, as formulated by Schroeder (1998) that:

the . . . relation between an emergent property of a whole and the properties of its parts is . . . one of non-explanatoriness.

This idea is an old one, as it was put forward by the British emergentists such as Mill (1872), Alexander (1920), Broad (1925), and to some extent by Morgan (1923), and has also been discussed in the more recent literature (see, for example, Wimsatt, 1972, and McLaughlin, 1992). In other words, the emergent property of the whole is inexplicable, i.e., non-deducible from the properties of the parts.

The opposite to "strong emergence" is "weak emergence," a point of view that more pragmatically asserts that the relationship between the whole and the parts may not be established because of technical difficulties, such as the lack of computational power or insufficient progress of our skills. Atmanspacher and Bishop (2002) discuss at length this point.

The view that emergent properties of molecules cannot be explained as a matter of principle on the basis of the components is opposed by several scientists, who argue that this is tantamount to assuming that a mysterious force of some non-defined nature is at work. In fact, strong emergence may sound like a kind of vitalistic principle – and this is something we do not necessarily want to reintegrate in science. For example, Bedau (1997) writes:

. . . to judge from the available evidence, strong emergence is one mystery we don't need.

Personally, I believe that the discrimination between a matter of principle (strong emergence), and a matter of practical difficulty (weak emergence), is not always possible – and perhaps it does not always make sense. Take the case of myoglobin, with its 143 amino-acid residues. Can the properties of myoglobin be predicted on the basis of the properties of the twenty amino acids?

There are 20^{143} possibilities of making a chain with 143 residues. In principle, the folding and then the binding properties of each of these different sequences could be calculated and therefore predicted, but this would take billions of years – aside from the fact that we really do not know yet how to make these calculations. Therefore, it is not simply a matter of time, but also a matter of our intellective power. At this point, is the impossibility of predicting the properties of myoglobin a matter of principle (strong emergence), or is instead a matter of practical impossibility (weak emergence)? I would maintain that it is not possible to make such a discrimination. Thus, more often than not, weak emergence is not distinguishable from strong emergence. I believe that in this way we can also accept the notion of strong emergence, if it is made clear that this does not imply the use of mysterious forces, but simply reflects the limits of our capability.

Thus, going back to our rose: at each level of increasing complexity on going from atoms/molecules to the entire organism, there are novel properties that cannot be described in terms of the lower constituents, and therefore the properties of a rose, or any other living organism, cannot be interpreted on the basis of atoms and molecules. This is so regardless of whether it is a matter of principle or a practical difficulty. Emergence really makes the difference between the rather primitive reductionistic interpretation and a more holistic view of reality.

Downward causation

It is generally accepted that the development of emergent properties, which is an upward (or bottom-up) causality, is attended by a downward – or top-down – causality stream. This means that the higher hierarchic level affects the properties of the lower components, as reflected by Schröder (1998):

Downward causation is the influence the relatedness of the parts of a system has on the behaviour of the parts . . . it is not the influence of a macro-property itself, but of that which gives rise to the macro-property, viz., the new relatedness of the parts.

There is an on-going discussion in philosophical literature on the relation between emergence and downward causation – also called macro-determinism – see for example Bedau (1997), Schröder (1998), and Thompson and Varela (2001). It is generally assumed that emergence and downward causation take place simultaneously; in particular, Thompson and Varela (2001) like to combine the occurrence of upward causation (emergence) with downward causation by using the notion of cyclic causality. Evan Thompson actually now likes to use the term "reciprocal causation" (personal communication, 2004), thus eliminating the uni-directionality of the word "downward".

Generally, the point can be made that molecular sciences and chemistry in particular offer very clear examples of downward causation, as defined above. In the trivial chemical example of the formation of water from the two gaseous components oxygen and hydrogen, the emergence of the water molecule profoundly affects the properties of both hydrogen and oxygen (due to binding orbitals). Likewise, the electronic orbitals of carbon atoms, and those of oxygen and hydrogen, are changed when the molecule of benzene is formed. In chemistry, relatedness means interaction, and any form of chemical interaction modifies the properties of the components.

It is important to emphasize that this kind of chemical relatedness does not need to assume any special effect other than the normal laws of chemistry and physics. As already mentioned, special effects are often invoked in the literature – particularly in the old "British emergentism." This has been one of the reasons why emergentism in general has been criticized, and once special forces are eliminated from the picture, then this criticism immediately loses validity, as pointed out by Schröder (1998) in his criticism of McLaughlin (1992).

Of course molecular-science examples are not the only ones to show the effect of downward causation. This is so for all aspects of emergence. Consider the progression of hierarchic levels that go from the individuals to the family to the tribe to the nation: it is clear that once the individuals have a family, the rules of the family affect and change the properties of the individuals, and so on.

Emergence and non-linearity

In the previous chapter self-organization was shown to have a very strong component in the domain of physics of non-linear systems, and actually some examples of self-organization in non-equilibrium systems have already been given, such as the beehive structure in silicon oil subjected to thermal gradient. This should be repeated at the level of emergence.

As already mentioned, the theoretical background of this aspect of emergence can be traced to the introduction of the dissipative structures by Prigogine. A dissipative structure in these terms is an open system that is in itself far from equilibrium, maintaining, however, a form of stability. In a pendulum, dissipation is caused by friction, which decreases the speed of the pendulum and eventually brings it to a stand-still. However, if energy is continuously provided to the system, the constant structure is maintained through a flow of energy. In more complex systems, depending on the initial conditions and fluctuations of the energy flow, the system in its dynamic behavior may encounter a point of instability – the bifurcation point – at which it can branch off with the emergence of new forms of structure and properties. According to Capra (2002):

The spontaneous emergence of order at critical points of instability is one of the most important concepts of the new understanding of life. It is technically known as self-organization and is often referred to simply as "emergence". It has been recognized as the dynamic origin of development, learning and evolution.

From this assertion the equivalence between emergence and self-organization may be recognised, which I have kept distinct for heuristic reasons. Capra then goes on to discuss the implementation of this kind of emergence not only in biological structures, but also in social systems (management, information, language), as well as in psychology. Again, the emphasis is on the intrinsic instability of the system, and as such it is useful to recall a citation by Prigogine (1997):

Once instability is included, the meaning of the laws of nature . . . change radically, for they now express possibilities of probabilities.

Because of this, as Christidis puts it (2002), matter acquires new properties when estranged from equilibrium, namely when fluctuations and instabilities are dominant. For a series of examples in biology, chemistry, and engineering, see Strogatz (1994).

How do we go from these theoretical concepts to living structures? Certainly living cells can be seen as open systems, and in fact in this case the link with the above thermodynamic relations is particularly interesting: the second law of thermodynamics states that the entropy of a closed system can only increase, or remain constant, in which case the system has reached a state of equilibrium. The cell does not proceed towards equilibrium just because the system is not closed; rather, it is an open system with a constant flux of energy during its life cycle. Autopoiesis, as will be seen shortly, can also be viewed as an expression of the same concept. To maintain this flux, dissipation is required, and we are thus back to the notion of dissipative structures. Actually, the term "dissipative" is often referred to structures that emerge as a result of self-organization and use dissipation to retain their organization.

Sidebox 6.1

Stuart Kauffman
Institute for Biocomplexity and Informatics
University of Calgary, Alberta, Canada
The sciences of complexity
The sciences of complexity appear to afford a countermeasure to the dominance of reductionistic science, which has held away for over three hundred years. It is not that Newton was unaware of the need to integrate the parts of a system into an understanding of the whole. Indeed, such an integration of his three laws of motion and law of universal gravitation gave us classical physics. Rather, in many of the fields mentioned above, the study of the collective behaviors of systems consisting of many interacting parts is gradually coming to the fore.

In physics, the late Danish physicist, Per Bak, and his colleagues Tang and Wiesenfeld startled the field in the late 1980s by producing a widely quoted paper on self-organized criticality. Bak and others applied this model widely – to the size distribution of earthquakes and the distribution of clusters of matter in the universe, to the size distribution of extinction events in the biological record. Self-organized criticality has been tested and succeeds in some cases, not in others. More important is the spirit of the effort: find collective variables and predict their collective behaviors, largely independent of the details of the individual parts.

In chemistry, one area that has received outstanding attention is that of RNA folding, shape, and evolution. Peter Schuster, Walter Fontana, Peter Stadler and their colleagues have made major contributions. Among the concepts here are "energy landscapes" for computer-folded models of RNA molecules, the evolution of model RNA sequences over these landscapes in sequence space, the folded shapes of model RNA sequences, and the existence of connected "neutral" pathways across sequence space among model RNA molecules that fold to the same shape.

Biology is emerging as a core field for the sciences of complexity. We have entered the post-genomic era. For fifty years, molecular biologists have focused brilliantly on the examination of specific genes, their "upstream" regulatory "*cis*" sites, the transcription factors binding such sites and the regulation of the activities of specific genes. It is now becoming abundantly clear that understanding the integrated holistic behavior of such networks is the next overwhelming challenge. One approach, begun by this author, is an ensemble approach in which one learns all the constraints present in real genetic networks, such as the number of molecular inputs that regulate genes and the distribution of that number across genes, the number of genes regulated by given genes, the control rules governing the activities of specific genes as a function of their inputs, the distribution of sizes of feedback loops, and so on.

Other approaches to genetic networks include study of small circuits with either differential equations or stochastic differential equations. The use of stochastic equations emphasizes the point that noise is a central factor in the dynamics. This is of conceptual importance as well as practical importance. In all the families of models studied, the non-linear dynamical systems typically exhibit a number of dynamical attractors. These are subregions of the system's state space to which the system flows and in which it thereafter remains. A plausible interpretation is that these attractors correspond to the cell types of the organism. However, in the presence of noise, attractors can be destabilized.

Progress in understanding genetic networks also includes attempts to solve the inverse problem. It is now possible to study the level of expression of some ten thousand genes simultaneously by obtaining RNA sequences from a set of cells. If the population of cells is differentiating, the patterns of gene activities change over time. Thus a "movie" can be constructed showing the waxing and waning of abundances of RNA sequences using gene chip arrays. An industry is growing that attempts to derive features of cells and circuits from such arrays. In one line of work, the inverse problem consists in viewing the movie and deducing the circuitry and logic driving the behavior of the genes.

In a variety of ways, the sciences of complexity are coming to bear on economics. For example, an agent-based model of a stock market has been constructed and it has shown that both fundamental pricing behavior and speculative bubbles can occur.

Another just emerging area of the converging of the sciences of complexity and standard economic theory is in the area of economic growth. Models in complexity theory have shown evidence that the rate of growth of an economy can be linked to the total diversity of goods and services in the economy, which are able to recombine combinatorially to create novel goods and services.

In computer science, advances have been made on several forefronts. These include the study by Crutchfield and colleagues of the emergence of pseudo-particles in cellular automata, where the particle interactions can carry out computations.

In management science, use of rugged landscape models has been brought to bear on questions concerning the adaptability of organizations, optimal organizational structure, the co-evolution of organizations, and further topics.

In summary, the sciences of complexity study the emergent collective behaviors of systems with many heterogeneous interacting parts. Those collective behaviors can now be studied with the computers that have become available. Computers will become more powerful. In a variety of scientific areas, the need to study the integrated behaviors of complex systems is becoming utterly apparent. This new body of science is science, not fad. It will grow, like all science, by stumbling forward, making mistakes, but with moments of real progress. That progress will complement the power of reductionism in leading us to a deeper understanding of myriad phenomena. Some of us, myself included, hope that general laws will emerge. That tale remains to be told.

As highlighted in Sidebox 6.1, Kauffman also stresses the link between biological systems and non-linear dynamic systems. This is a good introduction to the next section, which concerns emergence in some more complex biological systems.

Life as an emergent property

I have mentioned that most physicists would not consider the aromatic character of benzene or the cooperative behavior of hemoglobin as typical cases of emergence. Also, for most biologists, emergent properties become interesting only at the level of the complexity of the living systems; when dealing namely with tissues, organs, biological autonomy, self-reproduction, the behavior of social insects, and the like.

Biology offers innumerable examples where the increase of complexity is attended by the onset of sophisticated emergent properties. In the previous chapter the beehive-like structure of silicon fluids was mentioned; if we now consider a real honeycomb, each bee appears to behave as an independent element, acting apparently on its own account, but the whole population of bees gives a highly sophisticated collective emergent structure. The same can be said for an anthill or for any other type of social insects; in the previous chapter the pattern of flying birds was given as an example of "swarm intelligence."[1] Another very interesting phenomenon, that belongs to self-organization of "simpler" living systems, is the "quorum sensing." This is a cell density-dependent signalling mechanism used by many species of bacteria (Funqua *et al.*, 2001). Above a certain threshold of cell concentration, new collective properties appear in the colony, which are not present at lower cell density. Quorum sensing controls several important functions in different bacterial species, including the production of virulence factors and biofilm formation, as well as the capability of colonizing higher organisms, for example in *Pseudomonas aeruginosa*, as well as the appearance of bioluminescence in *Vibrio fischeri* (Miller and Bassler, 2001, Smith and Iglewski, 2003). It is basically an intercellular signalling, based on the production of special molecules, that activate genes, which in turn produce special proteins. There are already studies referring to the semantic and syntaxes of such a language (Ben Jacob *et al.*, 2004), as well as to interkingdom signalling (Shiner *et al.*, 2005). I find particularly fascinating the onset of bioluminescence in *V. fischeri* above a certain threshold of cell density, used as a recognition mechanism for members of the same species, which favor grouping and pairing. Strangely enough, this phenomenon is not usually considered in the field of emergence or complexity, however it is again the onset of collective properties due – in this case – to a clear chemical signal.

In all these cases, regular patterns of self-organization are realized without any externally imposed rules. There are complex self-organization patterns, but localized organization centers cannot be established anywhere.

In this regard, I would like to cite Francisco Varela in one of his last interviews before his death (in Poerksen, 2004):

Consider, for example, a colony of ants. It is perfectly clear that the local rules manifest themselves in the interaction of innumerable individual ants. At the same time, it is equally clear that the whole anthill, on a global level, has an identity of its own . . . We can now ask ourselves where this insect colony is located. Where is it? If you stick your hand into the anthill, you will only be able to grasp a number of ants, i.e., the incorporation of local rules. Furthermore, you will realize that a central control unit cannot be localised anywhere because it does not have an independent identity but a relational one. The ants exist as such but their mutual relations produce an emergent entity that is quite real and amenable to direct experience. This mode of existence was unknown before: on the one hand, we perceive a compact identity, on the other, we recognize that it has no determinable substance, no localisable core.

Actually, this concept of self-organization and emergent properties as a collective ensemble, without an organized localized centre, is nowadays under scrutiny in cognitive sciences: several scientists in the area would now agree that this

---

[1] For a more detailed view of swarm intelligence, see http://www.swarm.org.

notion of "I" is an emergent property arising from the simultaneous juxtaposition of feelings, memory, thoughts, remembrances . . . , so that the "I" is not localized somewhere, but it is rather an organized pattern without a centre. In order to make clear the analogy between this last concept and the organization/emergence in the case of social insects, let's again cite Varela, in the same reference as before:

This is one of the key ideas, and a stroke of genius in today's cognitive science. There are the different functions and components that combine and together produce a transient, non-localisable, relationally formed self, which nevertheless manifests itself as a perceivable entity. . . . we will never discover a neuron, a soul, or some core essence that constitutes the emergent self of Francisco Varela or some other person.

For more on cognitive science about this and analogous concepts, the reader is referred to the books by Damasio (1999), Varela (1999), and le Doux (2002).

Concluding remarks

There a few points that are worthwhile mentioning to conclude this chapter on emergence. One is the importance of the notion of non-predictability. The reason why this is important lies in the fact that novel, unpredicted properties can arise from the constitution of complexity. In other words, the fact that we cannot foresee novel emergent properties also means – and this is an exciting view – that there might be a vast arsenal of unforeseeable properties that may arise from the intelligent or serendipitous assemblage of components.

A second, unrelated point is the observation that the main concepts in the field of emergence become clear and simple, when using biological or chemical examples. This represents an important contribution of molecular science to the field of philosophy, a contribution – as already noted – not taken duly into account in the chemical literature and in chemistry textbooks. It would be advisable to introduce the notion of emergence in college books, so as to provide young chemistry and biology students with a broader perspective.

The third point goes back to the question between the epistemic and ontic views of philosophy of science. It has been mentioned a few times that the simple examples presented in this chapter are relative to an epistemic approach, namely to a reality in terms of descriptive terms, rather than in terms of "things as they really are" (the ontological view). The necessary context-independence of the ontic approach precludes its applicability to concrete cases. Conversely, as Primas (1998) states,

. . . the epistemic description contains a non-removable reference to the observing tools, the referent of such derived theories is the empirical reality.

This brings about an interesting relation between the ontic and the epistemic approach (Primas, 1998):

... an epistemic state of a derived theory refers to our partial knowledge of the ontic state of the fundamental theory.

In other words, a context-dependent theory still reflects some aspects of the independent reality.

Primas uses this argument to introduce the notion of contextual ontology, which refers to emergent properties arising from hidden features of the independent reality. This permits a clear view of the relation between contextual and fundamental theories, and also a generalization that is relevant for philosophy of science at large:

Only if we maintain multiple sets of contextual ontologies, we can tolerate the coexistence of complementary views in our experience of reality. While an independent reality itself is directly inaccessible, the numerous inequivalent contextual descriptions allow us to get deeper insight into the structure of independent reality (Primas, 1998).

Having indulged in philosophy, we can go back to biology. An important point about emergence is that life itself can be seen as an emergent property.

The single-cell components, such as DNA, proteins, sugars, vitamins, lipids, etc., or even the cellular organelles such as vesicles, ribosomes, etc., are per se inanimate substances. From these non-living structures, cellular life arises once the space/time organization is defined. The consideration that life is an emergent property gives the notion of emergence a particular significance. No vitalistic principle, no transcendental force, is invoked to arrive at life – and this, as mentioned already, has two consequences: (i) life, at least in principle, can be explained in terms of molecular components and their interactions; (ii) it is conceiveable to make some simple forms of life in the laboratory.

We have looked previously at a rose, and claimed that one would learn nothing about a rose by saying that it is composed solely of atoms and molecules. A better approach to the essence of the rose, would be to describe at least the various levels of hierarchic structural complexity and the corresponding emergent properties – up to the various cells and cell organelles, up to the different tissues; and then add possibly the history of biological evolution. This is certainly a more complete view of a rose. It is a departure from the simplistic reductionistic approach to see all in terms of atoms – but is it enough to catch the essence of a rose?

Most of the cognitive scientists mentioned above would add that what is still missing is the "observer" – the one who really gives meaning to the rose in terms of history, literature, poetry, . . . Obviously the notion of a rose is different depending on whether the observer is a Westerner educated in romantic literature, or an Eskimo who has never seen a rose. Here is where the notion of emergence may become co-emergence between the object and the observer throughout his/her consciousness. This is quite a difficult, complex subject, which we will return to in Chapter 8, which deals with autopoiesis and cognition.

Questions for the reader

1. Emergence gives rise to novel properties. Do you accept the idea that in the future unimaginable novel properties will emerge from the study of new composite materials or new synthetic complex systems?
2. Do you accept the idea that human consciousness is an emergent property of a particular neuronal and physical human construct?
3. After reading this chapter, do you adhere more to the view of "strong emergence" – or "weak emergence?"
4. You have extracted the nucleus from an oocyte. You have a nucleus and the rest of the cell. Neither of these two components is alive. Now you insert again the nucleus into the cell – or, in this respect – into a different enucleated cell (as in the cloning experiments). You get a living cell. Can this be taken as a demonstration for the emergence of life, to namely that life can be generated out of non-living components?

7

Self-replication and self-reproduction

Introduction

Let's start with a semantic note about the two terms self-replication and self-reproduction. Although often treated synonymously, they are not equivalent. There has been a discrimination between the two in the literature for some time, the clearest proponent being Dyson (1985). In fact, self-replication (the word comes from the Latin *replica*) means a faithful molecular copy, while self-reproduction refers rather to a statistical process of making very similar things. Thus, cells self-reproduce, while molecules self-replicate. It is not simply a semantic question, as according to Dyson (Dyson, 1985), self-reproduction processes, being less precise, in the early evolution most likely preceded self-replication processes, which require more complex control and editing.

Self-reproduction is rightly seen as the main motor for the development of life on Earth. Life unfolds in time by repetition patterns: one family of dolphins gives rise to a new family of dolphins, one family of roses gives rise to a new family of roses. This is a subtle, invisible thread that links the present beings to all those that have existed before and to all those who will come. Looking back, everything condenses into the original cellular family, the whimsical LUCA (last universal common ancestor); this, by the way, should not be confused with the origin-of-life protocell, as this is a much older event.

Self-replication and non-linearity

There are some good reasons why, in the field of the origin of life, the emergence of self-reproduction is of such paramount importance. Before the implementation of self-replication, any interesting structure that might have originated in the pre-biotic scenario would have decayed due to degenerative processes, and would have disappeared leaving no trace. Instead, with self-reproduction (as soon as the rate of

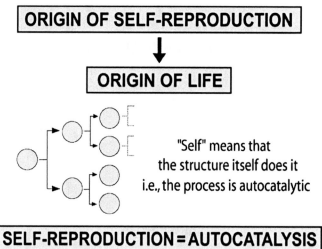

ORIGIN OF SELF-REPRODUCTION

↓

ORIGIN OF LIFE

"Self" means that
the structure itself does it
i.e., the process is autocatalytic

SELF-REPRODUCTION = AUTOCATALYSIS

If the structure carries information, then we have
⟹ **REPRODUCTION** & **INFORMATION** at the same time

If during self reproduction also structural (→ function)
changes occur, we also have **EVOLUTION**

Novel catalysis

↳ everything else...

Figure 7.1 Qualitative scheme of the relation between self-replication and the origin of life.

self-reproduction is larger than the rate of decay), an increase in concentration of this structure would be possible.

Also, if the self-replicating structure were to possess chemical information, and in addition the capability to mutate, then self-reproduction, information and evolution would occur all at the same time; something already very close to life (see Figure 7.1). It is then clear why the search for self-reproduction mechanisms is the holy grail of the research on the origin of life.

Shreion Lifson (1997) utilized a nice arithmetic to illustrate the power of the autocatalytic self-replication. He took the example of a normal hetero-catalytic process that makes one molecule of B from A at the rate of one per second. Then, it would require 6×10^{23} s to make one mole of B. If instead there is an autocatalytic process by which B gives rises to 2B, and 2B give rise to 4B, and 4B to 8B, and so on, it requires only 79 s to make one mole of B.

The dramatic power of non-linear growth was already known in ancient times. You may have come across the story of the Chinese Emperor who – according to one particular version – played chess with a concubine. The smart lady had

Figure 7.2 The power of non-linear growth. Put a grain of rice in the first square, two in the second, four in the third, eight in the fourth . . . and calculate how many kilograms of rice you have at the end of the chess board!

requested, should she be victorious, only a little bit of rice: one grain of rice in the first square of the chess board, two grains in the second, four in the third, and so on. The emperor nodded scornfully; he lost, and discovered that he didn't have enough rice in his kingdom to pay his debt (see Figure 7.2); and the same unproven version continues that this is how a concubine became empress of China and the emperor became a monk.

Back to more modern times, it may be recalled that the intellectual history of self-replication could go back to von Neumann and his cellular automata (von Neumann and Burks, 1966). (It is interesting that Alan Turing, who was mentioned in Chapter 6, completed a Ph.D. dissertation in 1934–38 under von Neumann's supervision). Von Neumann, aside from working in the Manhattan Project to develop the first atomic weapons, worked in quantum mechanics and devised the architecture used in most non-parallel-processing computers. Of most importance to us here, he created the field of cellular automata (von Neumann and Burks, 1966) without computers, constructing theoretically the first examples of self-replicating automata. Besides being a theoretician, von Neumann had a practical mind and his interest in self-replicating devices was also to show that practical operations such as mining might be succesfully accomplished through the use of self-replicating machines.

The names of von Neumann and Turing lead into the field of artificial life, a subject that falls outside the scope at this book. Freitas and Merkle (2004) discuss the subject beginning from von Neumann's cellular automation model, and usefully describe a large series of self-replication models presented in the artificial life literature.

Myths and realities of self-replication

The arithmetic expressed in Lifson's previous calculation, as impressive as it is, may give misleading conclusions. For example, one may be induced to think that one single molecule can do the whole trick: arise by chance and begin to self-replicate, giving birth to a vast family of molecular progeny. As already mentioned with regard to the RNA world, this idea – despite the early warning of Joyce and Orgel (1993) about the molecular biologists' dream – is still in the backyard of the RNA world, can be found in several college books, and it is even easily spotted in the specialized literature.

Obviously, with one single molecule no successful chemistry can be achieved (although it may be different with one DNA molecule in a living cell). In normal wet chemistry, in order to self-replicate, the replicator A must bind to another molecule A. It takes at least two molecules to make an active complex A–A, capable of starting a replication mechanism. In addition, in the case of nucleic acids one needs the four nucleotides (or whatever monomers are implied as substrates).

The need to form the A–A complex from two A molecules is a severe constraint. First of all, in order to make an appreciable concentration of this complex, there must be a significant amount of A, so as to overcome the effect of diffusion; and the real difficulty arises when spontaneous decay is introduced. If the concentration of A is low enough, the population will decline no matter how large the growth rate is.

In this regard, let us draw some lessons from contemporary RNA replication: RNA self-replicase does not exist in nature, the actual concentrations of $Q\beta$ replicase and template-RNA in a single cell may be considered, and compare with in vitro experiments (Szathmáry and Luisi, unpublished data). Based on the smallest dimension of a bacterium, a minimal concentration of c. 10 nM can be calculated for RNA in vivo.

For in vitro kinetics, in typical experiments, nucleotides are provided in 0.2–0.3 mM concentrations (Biebricher *et al.*, 1981), while the enzyme replicase is in the range of 80–140 nM (still about 10^{12} enzyme molecules per reaction sample!). Under such conditions the duplication time of molecules (about 120 nucleotide long) is about 2.3 hours, and becomes months or years when reagent concentrations are in the picomolar range and below. It may be argued that reaction times of months or years are not prohibitive in a prebiotic scenario; but here again is the problem of the decay rate of the replicator. Wong has calculated that for a genome consisting of three genes, the replication time must be shorter than 8.6 years (Wong and Xue, 2002). This seems to be a relaxed condition, but in fact it is not, once the actual concentrations are taken into account. In fact, it can be calculated (Szathmáry and Luisi, unpublished data), by using the in vitro RNA replication data, that the survival limit for the replicators must be in the nanomolar range of monomer concentration! This means, NB, the presence of c. 10^{17} identical copies of the replicator!

Aside from these calculations, and going back to the basic question of RNA self-replication, the important message from these simple calculations is this: that for a process of RNA replication several billions of identical RNA molecules and reasonably high monomer concentrations are needed. This should serve to abandon the naïve idea that one single RNA molecule can start life. In turn, on accepting the idea that a significant concentration of RNA is necessary, the conclusion must be that this amount of RNA must come from an intense cellular or proto-cellular metabolic activity. The use of vesicles or surfaces may alleviate the problem of concentration, but only to a certain degree, and certainly not to the point of eliminating this necessary prior metabolism.

From clay even less help can be expected: RNA and/or ribozymes in clay are being investigated, and certainly the rigid support of clay may have beneficial effects on the conformational stability of RNA. However, it is not clear what the possible biological meaning of such experiments for the origin of life is. Having elucidated the highly sophisticated structure of RNA, one does not want to go back to clay chemistry.

To some extent, this message has already been accepted in the literature, as the quest for the pristine RNA world has been shifted into a pre-RNA world. In a recent paper, Orgel (2003) appears to accept the view that the action of amino acids might have possibily formed the basis of this pre-RNA world.

Self-replicating, enzyme-free chemical systems

When considering self-replication, one immediately thinks of DNA, even though obviously DNA alone is not capable of replication: this process is only made possible by a large family of enzymes in a living cell. Likewise, a virus per se is not capable of self-reproduction, the reproduction is made by the living host cell. However, the double-strand nature of DNA is the most ingenious device for self-replication, as each of the two strands contains the information for making a complementary one. Figure 7.3 illustrates the well-known semi-conservative mechanism of DNA replication, which is the basis of the meiosis and mitosis processes in cell reproduction.

Are there other types of self-replication in nature, possibly based on a quite different mechanism? There are not many, but there is a famous case: the formose reaction, described in 1861, and based on a reaction cycle of formaldehyde. This reaction has already been mentioned in Chapter 3, on the subject of prebiotic

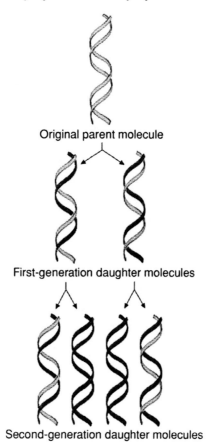

Original parent molecule

First-generation daughter molecules

Second-generation daughter molecules

Figure 7.3 The semi-conservative replication scheme of DNA.

chemistry (see also Figure 3.6); another pictorial rendering of this process is given in Figure 7.4: a cycle starts from one molecule of glycolaldehyde and ends up with two molecules of glycolaldehyde. In this way, after each cycle there is a doubling of the concentration of the final product. This reaction appears to be of limited meaning for making key molecules of life; and all other metabolic pathways and cycles that are found in biochemistry textbooks proceed – as in the case of the replication of DNA – with the help of many enzymes. However in a prebiotic scenario specific enzymes were not present and the question becomes whether, and to what extent, self-replication could have proceeded without enzymes.

Previous chapters have described the approach by Orgel and his group, with their template-directed synthesis of oligonucleotides. Also, we have seen that by using this technique self-replication was not achieved. Work describing oligonu-cleotide self-replication was published in 1986 by Gunter von Kiedrowski, now at the University of Bochum in Germany (von Kiedrowski, 1986; Sievers and

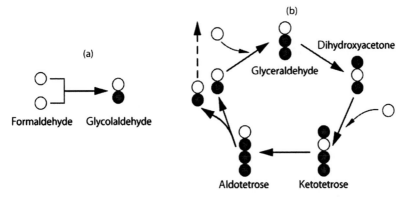

Figure 7.4 Schemae of the formose reaction: (a) spontaneous, slow formation of glycolaldehyde from formaldehyde; (b) after one cycle, one new molecule of glycolaldehyde is produced. The structural isomers of sugars are specified by the carbon skeleton and by the position of the carbonyl group (open circle). (Adapted, with some modifications, from Maynard Smith and Szathmáry, 1995.)

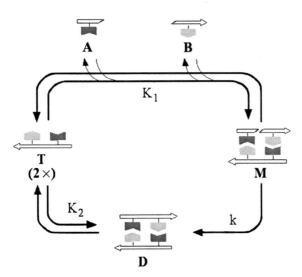

Figure 7.5 The enzyme-free self-replication scheme by von Kiedrowski (1986). The same scheme was used by Rebek *et al.* (1994) (see text) with non-nucleotidic molecules.

von Kiedrowski, 1994). The scheme is analogous to the semi-conservative replication of DNA and is illustrated in a more schematic form in Figure 7.5.

This approach utilized a particular hexanucleotide CCGCGG-p′ as a template: a 5′ terminally protected trideoxynucleotide 3′-phosphate d(Me-CCG-p), indicated as A, and a complementary 3′-protected trideoxynucleotide d(CGG-p′), indicated as B, were reacted in the presence of a condensing agent to yield the self-complementary hexadeoxynucleotide d(Me-CCG-CGG-p′), indicated as T, which is actually A–B. The main point of this synthesis was that the product T could act as a template for its own production, i.e., it could bind B to its moiety A and A to its moiety B.

The hexamer formation proceeds via the ter-molecular complex M, and the proximity between A and B in this complex facilitates their covalent linkage. Thus, once the complex D is dissociated, two T molecules are formed, and the autocatalytic self-replication process can start with the progression described above: two give four, four give eight, eight give sixteen, and so on.

Notice the particular features of this kind of oligonucleotide: the hexameric sequence is said to be self-complementary, since two identical molecules can form a duplex via Watson and Crick bases. It may also be noted from Figure 7.5 that two parallel pathways compete for the formation of the template T, namely the template-dependent, autocatalytic pathway, and the template-independent, non-autocatalytic one. This competition is the reason why the initial rate of the autocatalytic synthesis was found to be proportional to the square root of the template concentration – something that von Kiedrowski and colleagues called the square-root law of autocatalysis. As Burmeister (1998) put it:

a square root law is expected in the previously described cases, in which most of the template molecules remain in their double-helical complex (duplex) form, which leaves them in an "inactive" state. In other words, a square root law reflects the influence both of auto-catalysis and product inhibition.

Autocatalytic reactions that do not have this competition have a reaction order of one instead of a half; see also Zelinski and Orgel (1987).

One limit to this beautiful chemistry lies in the requirement of self-complementarity of the self-replicating sequences. The more general case of a template working by complementarity was also investigated by von Kiedrowski's group (Sievers *et al.*, 1994). As noted by Burmeister (1998), the underlying principle in this case is a cross-catalytic reaction in which one strand acts as a catalyst for the formation of the other strand. This is illustrated in Figure 7.6.

Julius Rebek and his group were also active at about the same time with enzyme-free self-replication of chemical structures. Unlike von Kiedrowski' group, he did not use nucleotides, but a replicator consisting of an adenosine derivative and a derivative of Kemp's acid (Rotello *et al.*, 1991; Rebek, 1994). See also Figure 7.7 for a self-replicating system not based on nucleic-acid chemistry. There are several variations of this scheme, which are not illustrated here – for reviews see Sievers *et al.*, 1994; Orgel, 1995.

Interesting elaborations of this original scheme were presented later by von Kiedrowski's group (Achilles and von Kiedrowski, 1993; von Kiedrowski, 1993;

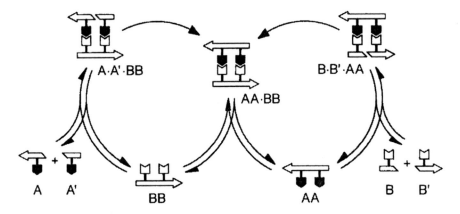

Figure 7.6 Minimal representation of a cross-catalytic self-replicating system. (Adapted from Burmeister, 1998.)

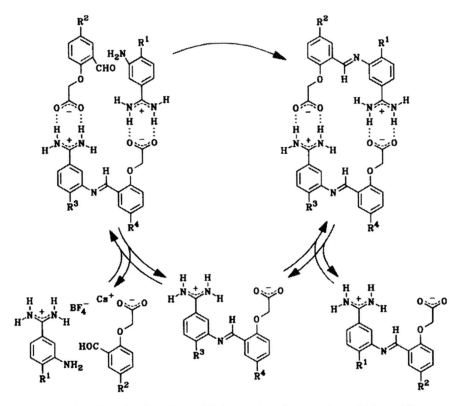

Figure 7.7 Self-replication of ammidinium carboxylate templates. (Adapted from Burmeister, 1998.)

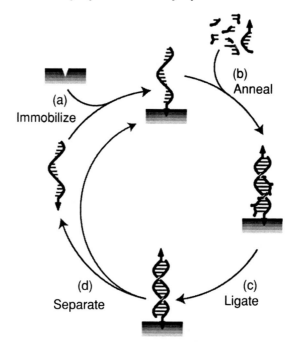

Figure 7.8 General scheme of the SPREAD procedure of self-replication. (a) A template is immobilized by an irreversible reaction with the surface of a solid support. (b) The template binds complementary fragments from solution. (c) The fragments are linked together by chemical ligation. (d) The copy is released, and re-immobilized at another part of the solid support to become a template for the next cycle of steps. Irreversible immobilization of template molecules is thus a means to overcome product inhibition. (Adapted from Luther *et al.*, 1998.)

Sievers *et al.*, 1994). Also, their more recent developments should be mentioned. By a technique called SPREAD (surface promoted replication and exponential amplification) the authors (Luther *et al.*, 1998) combined the original scheme of Figure 7.5 to an insoluble matrix, obtaining a quite complex reproduction scheme – see Figure 7.8. These reactions are very elegant and ingenious, they have however not been regarded as relevant for the origin of life.

A higher degree of complexity is offered by the elegant work by Nicolaou and coworkers (Li and Nicolau, 1994) (see Figure 7.9). In this case, the self-replication of DNA is based on a triple helix.

Of particular originality is the work by Ghadiri and coworkers, at the Scripps Institute, on self-replicating polypeptides (Lee *et al.*, 1996).[1] They showed that a

---

[1] Probably we at the ETH-Zürich were the first with an European collaborative research project (cost) to start work on self-replicating peptides, utilizing the same coiled-coil peptides as Reza Ghadiri's group. While we were playing with the physical characterization of these compounds, (Thomas *et al.*, 1995; 1997a, b; Wendt *et al.*, 1995), Ghadiri's brilliant work on self-replication appeared.

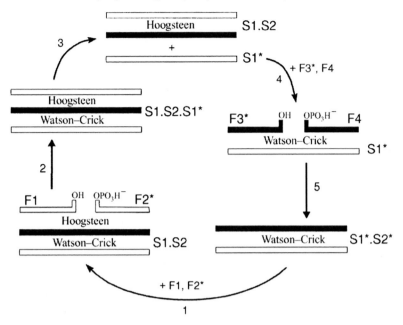

Figure 7.9 The triple-helix self-replication scheme by Li and Nicolaou, 1994 (adapted); S1 and S2 are the two strands, and F1 and F2 the two fragments. Purine strands are black; pyrimidine strands are white. For the precise meaning of the symbols in the figure, see the original paper.

32-residue alpha-helical peptide based on the leucine-zipper domain of the yeast transcription factor GCN4 can act autocatalytically, leading to its own synthesis by accelerating the thioester-promoted amide-bond condensation of 15- and 17-residue fragments in neutral, dilute aqueous solutions.

The original system is based on peptides that contain heptad repeats, where the first and fourth positions of the repeat are hydrophobic amino acids. Such sequences form α-helices, which assemble into coiled-coil structures, as represented in Figure 7.10. The principle is then the same as that used for von Kiedrowski's self-replicating nucleotides (von Kiedrowski 1986), in the sense that a full-length peptide template (having in this case 32–35 residues) directs the condensation of the two half-length peptide substrates.

The field of self-replicating peptides, initiated by Ghadiri's group (Lee *et al.*, 1996 and 1997) was then elaborated by Chmielewski and coworkers (Yao *et al.*, 1997 and 1998; Issac and Chmielewski, 2002; Li and Chmielewski, 2003) as well as by Mihara's group (Matsumura *et al.*, 2003; Takahashi and Mihara, 2004). Recent advances from Chmielewski's laboratory have eliminated the initial problems of slow release of the newly formed condensation product and have actually

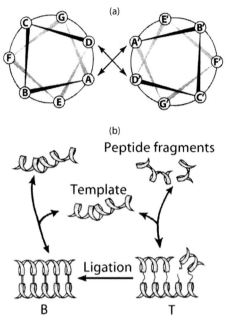

Figure 7.10 Coiled-coil structures during peptide condensation. (a) Two peptides containing heptad repeats (A to F and A′ to F′) able to form α-helices. Hydrophobic interactions (A–A′ and D–D′) lead to coiled-coil structures. (b) A full-length peptide (T) acts as a template forming a coiled-coil structure with peptide fragments and directing their condensation in other full-length peptides (B). (Adapted from (a) Paul and Joyce, 2004; (b) Ghosh and Chmielewski, 2004.)

approached exponential amplification. For reviews, see Paul and Joyce (2004), as well as Gosh and Chmielewski (2004).

Ghadiri's group also made the observation that a longer peptide actually works as a peptide ligase (Severin *et al.*, 1997): in fact, these authors showed that the 33-residue synthetic peptide, based on the coiled-coil structural motif, efficiently catalyzes the condensation of the two shorter peptide fragments. Depending on the substrates used, rate enhancements of tenfold to 4,100-fold over the background were observed. Furthermore, they extended and developed this work to include ecological systems consisting of hypercycles (Lee *et al.*, 1997). Starting with an array of 81 sequences of similar 32-residue coiled-coil peptides, the authors estimated the relative stability difference between all plausible coiled-coil ensembles and used this information to predict the auto- and cross-catalysis pathways and the resulting plausible network motif and connectivities.

Networks of self-replicating replicators are supposed to be the next step in complexity, and in fact, Ghadiri's and coworkers devised a complex hypercycle network of self-replicating and cross-replicating peptides (Lee *et al.*, 1997;

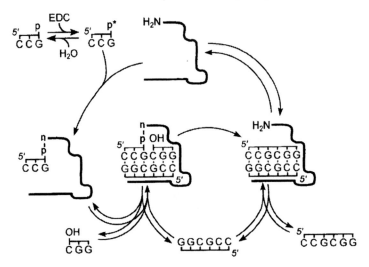

Figure 7.11 Principle of a minimal replicase. (Adapted from Maynard-Smith and Szathmáry, 1995. See also, with comments, Burmeister, 1998.)

Ashkenasy *et al.*, 2004). Whether or not such complexity can teach us anything about the origin of life and/or evolution, remains to be seen. One argument in favor is that such catalytic intercrossing networks may suggest a way by which the world of peptides and the world of nucleotides may have interacted with each other. Ellington and coworkers have taken a step in this direction by devising a system in which there is peptide-templated nucleic acid ligation with a RNA aptamer (Levy and Ellington, 2003). In this general framework the work by Mihara's group, who have incorporated nucleobase analogues in a self-replicating peptide (Matsumura *et al.*, 2003), should also be noted.

The general question arising in connection with Ghadiri's and Chmielewski's work, aside from its brilliancy and chemical ingenuity, is whether self-replication of peptides is per se relevant for the origin of life. Probably the answer is negative, but the very important message coming from these studies is a general one, the indication namely that autocatalytic self-replication processes are not mysterious, strange chemical pathways, but on the contrary, they enjoy a certain degree of generality in the world of chemistry.

One more step towards complexity

One of the dreams of present-day researchers is the construction of a minimal enzyme capable of making copies of oligonucleotides – a RNA polymerase. The ideal operation scheme is presented in the Figure 7.11. Here, the enzyme first covalently binds an activated trinucleotide, then a hexanucleotide can bind by

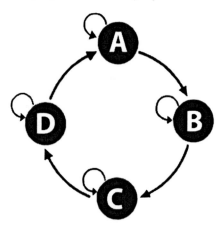

Figure 7.12 A simple rendering of an hypercycle. Each of the units A, B, C and D is a replicator. The rate of replication of each unit is an increasing function of the concentration of the unit immediately proceeding it. Thus the rate of replication of B is an increasing function of the concentration of A, and so on round the cycle. (Adapted from Maynard-Smith and Szathmáry, 1995.)

complementarity with the triplet, a second triplet can now bind, and the enzyme links together the two trinucleotides. Thus, two palindromic hexanucleotides are fabricated, and when their duplex divides up, the cycle can repeat itself – and each time, from one hexanucleotide, two are formed. Such an enzyme, or ribozyme, does not exist yet, although an approximation to it was described by Paul and Joyce (2002). This is a self-replicating system based on a RNA template that also functions as a ribozyme to catalyze its own replication – considered as the first example of a self-replicating system with a reaction order of unity.

Until now, single self-replication units have been considered, working with a circular logic of autocatalysis. Quite interesting and important is the next level of complexity, in which more self-replicative cycles interact and cooperate with each other. This gives rises to the notion of *hypercycle*, originally developed by Eigen and collaborators (Eigen, 1971; Eigen and Schuster, 1977; 1979), and schematized in Figure 7.12. The notion of hypercycle refers to self-replicating informational macromolecules and introduces into the mechanism the basic principle of Darwinian evolution. One important point is the notion of "quasi-species:" due to the ongoing production of mutant sequences, selection does not act on single sequences, but on mutational "clouds" of closely related sequences, referred to as quasi-species. The evolutionary success of a particular sequence depends not only on its own replication rate, but also on the replication rates of the mutant sequences it produces, and on the replication rates of the sequences of which it is a mutant. As a consequence, the sequence that replicates the fastest may even disappear completely in selection–mutation equilibrium, in favor of more slowly replicating sequences that are part of a quasi-species with a higher average growth rate. However, although the sequences that replicate more slowly cannot self-sustain, they are constantly replenished, as sequences that replicate faster mutate into them. Thus, at equilibrium, removal of slowly replicating sequences due to decay or out-flow is balanced, so that even relatively slowly replicating sequences can remain present in finite abundance (Eigen, 1971; Eigen and Schuster, 1979; Schuster and Swetina, 1988). The concentrations of reagents are mutually dependent, and the systems are coupled cyclically so that each unit helps its neighbor to replicate better.

The system is symbiotic, and is considered the basis of co-evolution; as such it has also been investigated by many authors from a theoretical point of view. An in-depth discussion is presented by Kauffman (1993).

Self-reproducing micelles and vesicles

So far, only the self-replication mechanisms of linear molecules have been described; it is now time to consider closed spherical structures, such as micelles and vesicles. Here, the term self-reproduction will be used rather than self-replication, because, as it will be seen, the population increase is generally based on statistical processes. The subject of micelles and vesicles self-reproduction is dealt with in other chapters in this book: a certain degree of repetition and/or mis-match is unavoidable.

The work was experimentally initiated in my laboratory in Zürich in the 1980s, at a time in which I was very much involved with reverse micelles – and in fact reverse micelles were the first spherically closed system that underwent self-reproduction.

Reverse micelles are small (1–2 nm in diameter), spherical surfactant aggregates built in an apolar solvent (usually referred to as oil), whereby the polar heads form a polar core that can contain water – the so-called water pool. The connection with autopoiesis is historically important, because it was with the collaboration with Francisco Varela that the work started (in fact it began as a theoretical paper – see Luisi and Varela, 1990). The idea was this: to induce a forced micro-compartmentalization of two reagents, A and B, which could react inside the boundary (and not outside) to yield as a product the very surfactant that builds the boundary (Figure 7.13). The product S would concentrate at the membrane interface, which increases its size. Since reverse micelles are usually thermodynamically stable in only one given dimension, this increase of the size-to-volume ratio would lead to more micelles. Thus the growth and multiplication would take place from within the structure of the spherically closed unit, be governed by the component production of the micellar structure itself, and therefore (as will be seen better in

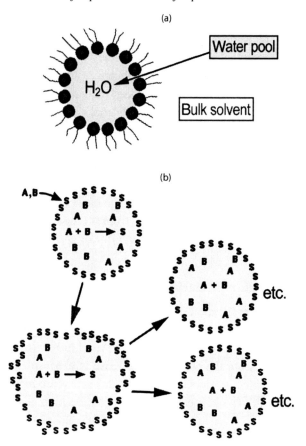

Figure 7.13 The first self-reproduction scheme conceived for reverse micelles. (a) A reverse micelle. (b) Two reagents, A and B, penetrate inside the water pool and react with each other inside the boundary, yielding the very surfactant S that makes the micelles. The S thus produced migrates to the boundary and induces growth and eventually multiplication of the micelle. (Adapted from Luisi and Varela, 1990).

the next chapter) we would be dealing with an autopoietic system, although in a very simple form. Note that this would also be an autocatalytic process: the more micelles are formed, the more the chemical process of making more micelles speeds up.

However, the experimental procedure that was applied had to be slightly different, as it is not easy to build a surfactant starting from two water-soluble moieties. Rather, one of the two should be lipophylic, as shown in the Figure 7.14.

A few systems of this kind were developed (Bachmann, 1990 and 1991), and this allowed the introduction of the notion of self-reproducing micelles. The

(a)

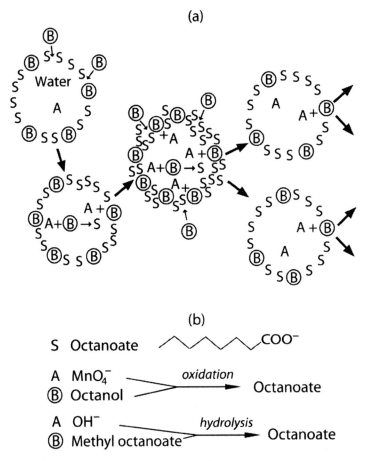

(b)

S Octanoate   /\/\/\/COO⁻

A MnO₄⁻ ——— *oxidation*
Ⓑ Octanol ———————→ Octanoate

A OH⁻ ——— *hydrolysis*
Ⓑ Methyl octanoate ———————→ Octanoate

Figure 7.14 Self-reproduction in reverse micellar systems: (a) The proposed mechanism of incorporation of the added precursor B, and the reaction A + B leading to the surfactant S. Ⓑ, Lipophylic cosurfactant (in excess in the bulk); A, localized in the micellar core; S, surfactant (A + Ⓑ → S). (b): Examples of some investigated chemical systems. (Adapted from Bachmann *et al.*, 1990 and 1991.)

compartmentation in reverse micelles has also been utilized to host the rather complex von Kiedrowski reaction of the self-replication of an hexanucleotide (Böhler *et al.*, 1993). In this case, while the hexanucleotide self-replicates according to the von Kiedrowski mechanism, the reverse micelles also undergo a self-reproduction – thus providing a system in which shell and core replication occur simultaneously (although not coupled to each other). Later, the procedure for self-reproduction of reverse micelles was developed into a procedure for aqueous micelles as well as for vesicles.

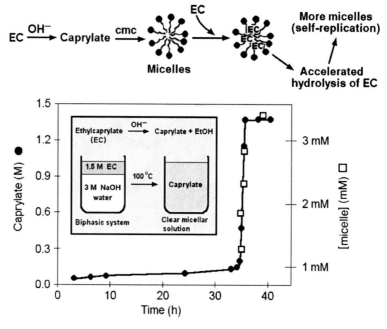

Figure 7.15 Autocatylatic self-reproduction of aqueous caprylate micelles. (Adapted, with some modifications, from Bachmann *et al.*, 1992.)

Let us recall the micellar aqueous system, as this procedure is actually the basic one. The chemistry is based on fatty acids, that build micelles in higher pH ranges and vesicles at pH *c.* 8.0–8.5 (Hargreaves and Deamer, 1978a). The interest in fatty acids lies also in the fact that they are considered possible candidates for the first prebiotic membranes, as will be seen later on. The experimental apparatus is particularly simple, also a reminder of a possible prebiotic situation: the water-insoluble ethyl caprylate is overlaid on an aqueous alkaline solution, so that at the macroscopic interphase there is an hydrolysis reaction that produces caprylate ions. The reaction is very slow, as shown in Figure 7.15, but eventually the critical micelle concentration (cmc) is reached in solution, and thus the first caprylate micelles are formed. Aqueous micelles can actually be seen as lipophylic spherical surfaces, to which the lipophylic ethyl caprylate (EC) avidly binds. The efficient molecular dispersion of EC on the micellar surface speeds up its hydrolysis, (a kind of physical micellar catalysis) and caprylate ions are rapidly formed. This results in the formation of more micelles. However, more micelles determine more binding of the water-insoluble EC, with the formation of more and more micelles: a typical autocatalytic behavior. The increase in micelle population was directly monitored by fluorescence quenching techniques, as already used in the case of the

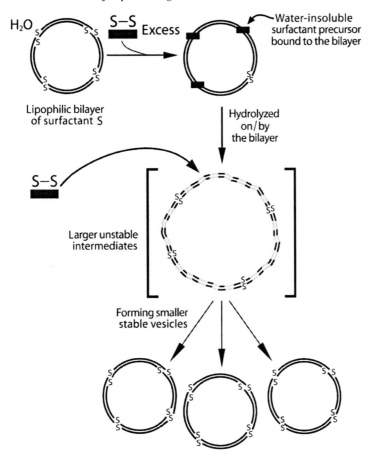

Figure 7.16 Self-reproduction scheme of vesicles; S–S represents a water-insoluble precursor that binds to the bilayer and is hydrolyzed *in situ*.

self-reproduction of reverse micelles (Bachmann *et al.*, 1992). Micelles could then be converted into vesicles – detected by electron microscopy – by lowering the pH by addition of CO_2.

This kind of process was reinvestigated later, among others by Micheau and his group (Buhse *et al.*, 1997; 1998). In the following years, this type of experiment was extended to vesicles. The experimental set up was similar, with a water-insoluble S–S precursor of the S-surfactant binding at the surface of the vesicles, as indicated in Figure 7.16. Again, the large hydrophobic surface of the vesicles accelerates the hydrolysis of S–S. This starts an autocatalytic self-reproduction process, as S remains in the membrane, which grows and eventually divides, giving rise to more aggregates.

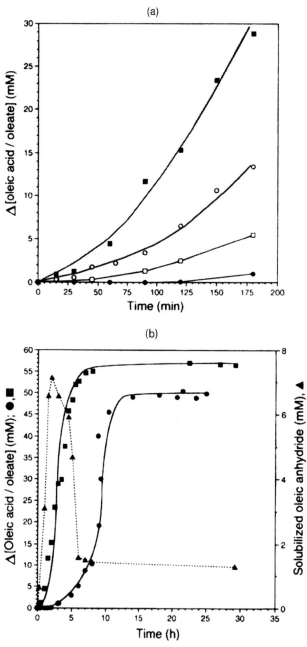

Figure 7.17 Hydrolysis of oleic anhydride catalyzed by spontaneously formed oleic acid vesicles at 40 °C, (a) during the first 3 h, and (b) during a long observation time. A vesicle suspension (10 ml in 0.2 M bicine buffer (pH 8.5)) was overlaid with 0.25 mmol oleic anhydride and 0.025 mmol oleic acid. The increase of the concentration of oleic acid/oleate is plotted as a function of reaction time. Initial concentration of oleic acid/oleate: 0 mM (•), 5 mM (□), 10 mM (○), 20 mM (■). For an initial oleic acid/oleate concentration of 20 mM, the concentration of oleic anhydride (▲) present in the vesicles during the reaction is also plotted (b, right axis). (From Walde *et al.*, 1994b.)

The time course of an actual experiment is shown in Figure 7.17, which shows the hydrolysis of oleic anhydride catalyzed by spontaneously formed oleate vesicles. Note the sigmoid behavior, typical of an autocatalytic process. The lag phase is due to the preliminary formation of vesicles, and in fact the length of the lag phase is shortened when already formed vesicles are pre-added, as shown in the figure. Some mechanistic details of these processes will be discussed in Chapter 10. In this work, an analysis of the number and size distribution of vesicles at the beginning and the end of the reaction was also performed by electron microscopy.

The sigmoidicity of the curves is strongly dependent on pH as well as on temperature, as shown in the case of caprylate vesicles in Figure 7.18. This phenomenon has not yet been studied in detail.

Self-reproduction studies were conducted on other fatty acid/anhydride systems, including attempts to correlate self-reproduction with stereospecificity by using chiral fatty acids. This was the case for the vesicles of 2-methyldodecanoic acid (**4**) (Morigaki *et al.*, 1997). The structures of the compounds and their reaction pathway are shown in Figure 7.19. The rate of hydrolysis of the chiral 2-methyldodecanoic anhydride (**8RR** and **8SS**), catalyzed by (**4R**) or (**4S**) vesicles, was studied with the aim of possibly combining exponential autocatalysis and enantioselectivity, i.e., to use the power of exponential growth to discriminate between two diastereomers. No significant effect of this kind was found. However, a marked difference in the behavior of homochiral and "racemic vesicles" was observed at 10 °C: the racemic vesicles separated out in gel-like form, whereas the homochiral ones remained stable and continued to self-reproduce during hydrolysis of the anhydride. Also, a significant influence of temperature on the hydrolysis rate and cooperativity was observed (Morigaki *et al.*, 1997).

In general, the mechanism of self-reproduction of micelles and vesicles can be considered an autopoietic mechanism, since growth and eventually division comes from within the structure itself. This point will be considered again in Chapter 8, on autopoiesis, where the mechanism of the self-reproduction process will also be discussed.

The term catalysis in this case is also a delicate point, as we are dealing with a kind of physical catalysis (due mainly to the large surface area of the surfactant aggregates) more than to a decrease of the activation energy of key reactions.

It is important to point out the main message of these experiments. This is that, by a very simple set-up, a spontaneous self-reproduction of spherical compartments can be obtained. Since such spherical compartments can be considered as models and/or precursors of biological cells, the hypothesis was put forward, (Bachmann *et al.*, 1992), that this autocatalytic self-reproduction process might have been of relevance for the origin of life.

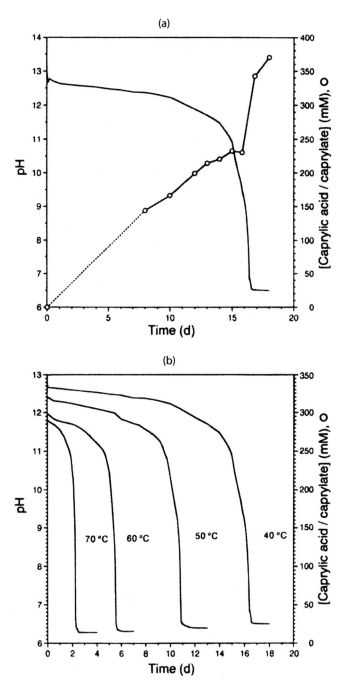

Figure 7.18 Hydrolysis of caprylic anhydride leading to the formation of self-reproducing caprylic acid vesicles. A mixture of 10 ml 0.265 M NaOH, 0.1 M NaCl, and 2.5 mmol of caprylic anhydride was incubated at a fixed temperature under slight stirring. (a) Change in pH and caprylic acid/caprylate concentration in the aqueous phase as a function of reaction time at 40 °C. (b) Variation of the reaction temperature. (From Walde *et al.*, 1994b.)

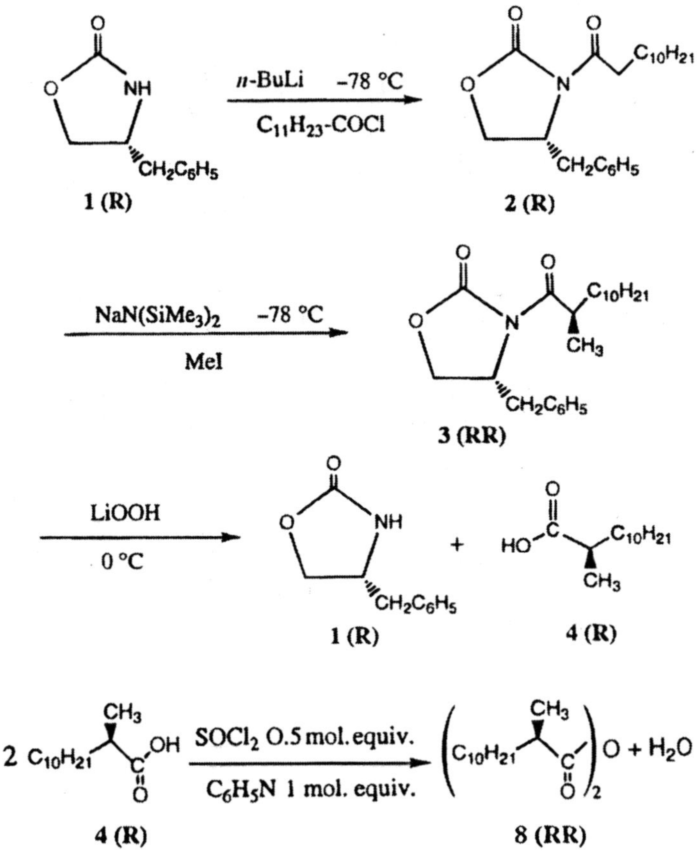

Figure 7.19 Pathway for the synthesis for 2-methyldodecanoic acid (**4**) and 2-methyldodecanoic anhydride (**8**), used for research on correlation between self-reproduction and stereospecificty. (Adapted from Morigaki *et al.*, 1997.)

In the various reviews on self-reproduction in recent years, practically no mention is made of such micelles or vesicle systems. The reason lies most probably in the bias of classic biochemical literature, according to which self-replication is tantamount to nucleic acid; systems lacking this are therefore deemed not to be relevant. In this particular regard, it is argued that self-reproduction of micelles and vesicles proceeds without transmission of information.

One counter-argument is presented by the group of Doron Lancet at the Weizmann Institute, who has been pushing for the notion of an informational content in liposome reproduction when starting from vesicles formed by a mixture of surfactant molecules. Sidebox 7.1, by Doron Lancet, gives an insight into this concept.

Sidebox 7.1

Doron Lancet

The Weizmann Institute of Science, Rehovot, Israel

Lipid-world "composomes" as early prebiotic replicators

How can non-covalent aggregates, made of molecules normally considered to be humble shell-formers, store and propagate information? The Graded Autocatalysis Replication Domain (GARD) model (see Segre and Lancet, 2000; Segre *et al.*, 2000; 2001) provides a possible, though unorthodox solution. A crucial focus of this model is the notion of compositional information. It is proposed that in early prebiotic evolution, molecular assemblies could assume a distinct individuality via the counts of different organic compounds within them. A useful relevant analogy is the propagation of epigenetic information in present-day cells. It is demonstrated mathematically that such compositions may encode Shannon information just like sequences. Prior to cell division, and in parallel with DNA replication, dividing cells must grow homeostatically, preserving the concentrations of thousands of components by de-novo metabolism-based synthesis. Only then may fission result in two identical progeny cells.

Detailed computer simulations of GARD assemblies (see Shenhav and Lancet, 2004) show that they are capable of a similar homeostatic growth, by selective absorption of compounds from the environment. The specific compositional states capable of homeostatic growth and information-preserving splits are called "composomes" (see Segre *et al.*, 2000 and Figure 7.20). Their behavior is inferred from numerical solutions of differential equations governed by rigorous kinetic and thermodynamic rules. These, in turn, underlie transitions from one composome to another, in a simple evolutionary progression. In this respect, composomal amphiphilic assemblies may justly be regarded as primitive protocells, capable of autopoietic reproduction and possessing life-like properties.

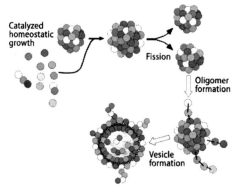

Figure 7.20 The "composomes", specific compositional states capable of homeostatic growth and information-preserving splits.

Concluding remarks

No longer than twenty years ago, self-replication was one of those mysterious processes considered the monopoly of living matter. The fact that we are now able to achieve it in the laboratory means that we understand self-replication and self-reproduction in terms of simple rules of chemistry. In turn, this means that we have proceeded a step further in the understanding of the mechanisms of life. Of course, this is just one step, but it shows that conceptual and experimental progress in the ladder of the transition to life is advancing.

We have also learned that self-replication is not a prerogative only of nucleic acids, but it can be shared by different kinds of chemical families; see the formose reaction, the self-replicating peptides, and the self-reproducing micelles and vesicles. The list should include the cellular automata and the corresponding devices of artificial life. Self-reproduction of vesicles and liposomes is important because it represents a model for cell reproduction.

I have brought into this chapter another line of criticism against the naïve version of the "prebiotic" RNA world, in particular about the assumption that self-replication and the corresponding molecular evolution processes may be sustained by one single molecule. Clearly, self-replication in a prebiotic scenario, in order to be chemically important, has to respect realistic concentrations and rate constants. It may be different in a fully fledged cell, once specialized enzymes and biochemical matrices have evolved – but this is a point of arrival and not of origination.

I would like to end this chapter with a citation by Dyson (1985), cited in turn by Lifson (1997), as it contains a flavor of criticism towards the emphasis on the importance of self-replication in the evolution of life, and a kind of "ante litteram" stand for system biology:

I have been trying to imagine a framework for the origin of life, guided by a personal philosophy which considers the primal characteristics of life to be homeostasis rather than replication, diversity rather than uniformity, the flexibility of the genome rather than the tyranny of the gene, the error tolerance of the whole rather than the precision of the parts (. . .). I hold the creativity of quasi-random complicated structures to be a more important driving force of evolution than the Darwinian competition of replicating monads.

There are probably nowadays more and more people who would take this stand; and this is probably good.

Questions for the reader

1. Suppose you find a colony of bacteria with a metabolic life but in which you are unable to measure any self-reproduction. Would you call them alive?
2. The above mentioned colony of bacteria without measurable self-reproduction would not be able to evolve. It would be life without evolution. Would that be acceptable to you?
3. What does your intuition say: did macromolecular self-replication systems come first in the origin of life; or should they be seen as the product of a mature cellular or proto-cellular metabolism?

8

Autopoiesis: the logic of cellular life

Introduction

We saw in Chapter 2 how the Green Man and the farmer, in collaboration, arrived at a descriptive definition of life which distinguished living from non-living. The farmer was ignorant of biology, otherwise he would have answered from the very beginning that all living elements are made up of cells – that this is the most discriminating factor. However, the Green Man would have then asked: "What is a cell and why is a cell living?"

In fact, the life of a cell is the starting point for the development of the ideas of autopoiesis (from the Greek *auto*, or self, and *poiesis*, or producing) developed by Maturana and Varela (Varela *et al.*, 1974; Maturana and Varela, 1980; Maturana and Varela, 1998). The aim of this chapter is to review the notion of autopoiesis and to present it in the context of present-day research in the life sciences. This will imply some addition to, and modification of, the original theory and also of a recent review of mine on which the first part of this outline is based (Luisi, 2003b).

This work is also prompted by the observation that cellular theories, such as autopoiesis, are once again attracting attention due to developments in the field of experimental cellular models. In this chapter, the notion of the *chemoton* developed by Tibor Ganti (Ganti, 2003) at about the same time as Maturana and Varela's autopoiesis will also briefly be considered, in order to see the analogies and differences between these two conceptually related viewpoints. Autopoiesis deals with the question "What is life?" and attempts to isolate and define, above and beyond the diversity of all living organisms, a common denominator that allows for discrimination between the living and the non-living. Autopoiesis is not concerned with the origin of life per se, or with the transition from the non-living to the living; rather, it is an analysis of the living as it is – here and now, as the authors say. Autopoiesis also deals with the various processes connected with life, such as

155

interaction with the environment, evolution, and cognition, and attempts to interpret these aspects in a coherent conceptual scheme. As a consequence, a series of epistemological concepts emerges from the analysis, part of which will be encountered in this chapter.

Historical background

Autopoiesis is embedded in a particular cultural background and therefore some historical information may be useful. When, as a student, Francisco Varela met Humberto Maturana in the 1960s at the University of Chile in Santiago, Maturana was already internationally known for his work on visual perception in frogs. This was the basis for the later work with Varela opposing representationalism in perception (see Maturana and Varela, 1998). The biological basis of cognition had always been an important item on Maturana's agenda (Maturana *et al.*, 1960) and this too was destined to influence significantly the interests and the later work of Varela. Varela left Chile in 1968 to pursue a Ph.D. in biology at Harvard, where he also had the opportunity to develop his interest in philosophy: Husserl, Heidegger, Piaget, and Merleau-Ponty were particularly important for his later work. Varela went back to the University of Chile in 1970. The term autopoiesis was used for the first time by the Santiago authors in 1971; by the end of that year Maturana and Varela had prepared a very long manuscript entitled "Autopoiesis: The Organization of Living Systems." As Varela describes (Varela, 2000), the manuscript was not well-received. It was rejected by the most important journals, and colleagues' responses were lukewarm. This was also a hard period for Varela because of the political situation in Chile: President Allende was assassinated and Varela, who had supported him, lost his job and had to leave the country in 1973.

Finally a paper on autopoiesis was submitted and later published in English (Varela *et al.*, 1974). The notion of autopoiesis was very slow to receive recognition. By the mid 1970s, however, some international meetings had used the term in their programs; books on autopoiesis appeared (Zeleny, 1977) and noteworthy biologists such as Lynn Margulis accepted autopoiesis as an integral part of the description of life (Margulis and Sagan, 1995). The notion of autopoiesis had a strong impact in the social sciences where the term "social autopoiesis" was coined (Luhmann, 1984; Mingers, 1992; 1995; 1997). In the 1990s, experimental chemical systems, on the basis of a previously elaborated theoretical scheme (Luisi and Varela, 1990), were developed in Zürich focusing on autopoiesis (Bachmann *et al.*, 1992; Walde *et al.*, 1994b; Luisi, 1996, and references therein).

Still, it cannot be said that the notion of autopoiesis is now familiar in mainstream science. The reason for this will be discussed later on, but it can be anticipated that this is partly due to the fact that autopoiesis theory is not centered on DNA, RNA, and

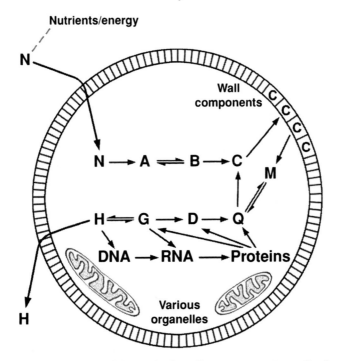

Figure 8.1 Schematization of the work of a cell as an open system. One important feature is the boundary, which is created by the internal network of reactions (a boundary of its own making). The network of reactions brings about a large series of transformations; however under homeostatic conditions all material that disappears is generated again by the internal machinery. Thus, the cell (and by inference, life) can be seen as a factory concerned with self-maintenance.

replication, and makes only minimal use of the term "information." Furthermore, the fact that it has been used extensively in the social sciences, and not always in a very rigorous way, has given to some the impression that it might even be tainted by a "new-age" flavor.

Basic autopoiesis

The autopoietic analysis of life is based on cellular life, the main argument for this being simply that there are no other forms of life on earth. We all know that even the simplest cells are extremely complex, encompassing hundreds of genes and other macromolecules. However, beyond this complexity, the question of what a cell really does, lends itself to a relatively simple answer. Consider Figure 8.1, which schematizes a cell. The first thing one observes is the boundary, a semi-permeable, spherical closed membrane that discriminates the cell from the medium. Here the term semi-permeable means that certain substances (nutrients and other chemicals) are able to penetrate in the interior, whereas most other chemicals cannot. This is a kind of chemical selection and chemical recognition – a notion that will be used later on in connection with the term "cognition."

The notion of boundary is, in fact, one central concept in the theory of autopoiesis. Inside the boundary of a cell, many reactions and correspondingly many chemical transformations occur. However, despite all these chemical processes, the cell always maintains its own identity during its homeostasis period. This is because the cell (under steady-state conditions and/or homeostasis) regenerates within its own boundary all those chemicals that are being destroyed or transformed, ATP, glucose, amino acids, proteins, etc.

The chain of processes occurring inside the boundary essentially serves the purpose of self-sustenance, or auto-maintenance. Of course, this takes place at the expense of nutrients and energy coming from the medium. As discussed earlier, the cell is a dissipative, open system.

From these simple, basic observations, Maturana and Varela (often referred to as the Santiago school) arrived at a characterization of living systems based on the autopoietic unit. An autopoietic unit is a system that is capable of sustaining itself due to an inner network of reactions that regenerate the system's components (Varela *et al.*, 1974; Maturana and Varela, 1980; Luisi, 1997; Maturana and Varela, 1998; Varela, 2000; Luisa *et al.*, 1996).

In other words, it can be said that an autopoietic system organizes the production of its own components, so that these components are continuously regenerated and can therefore maintain the very network process that produces them.

It is perhaps pertinent at this point to cite a recent definition by Maturana himself (in Poerksen, 2004):

When you regard a living system you always find a network of processes or molecules that interact in such a way as to produce the very network that produced them and that determine its boundary. Such a network I call autopoietic. Whenever you encounter a network whose operations eventually produce itself as a result, you are facing an autopoietic system. It produces itself. The system is open to the input of matter but closed with regard to the dynamics of the relations that generate it.

In this way, autopoiesis is capable of capturing the mechanism that generates the identity of the living. A graphic representation of these concepts reveals a circular logic, as shown in Figure 8.2. The components organize themselves (auto-organization) in a bounded system that produces the components that in turn produce the system . . . etc. In this way, the blueprint of life obeys a circular logic without an identified beginning or end, as pointed out by Maturana in the previous citation. Although the system is open from the physical point of view, from an epistemological perspective it has a logical operational closure (Varela, 2000; Maturana and

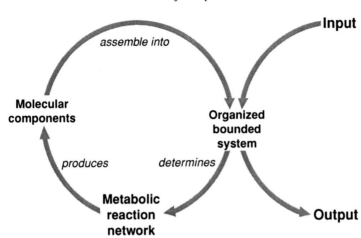

Figure 8.2 The cyclic logic of cellular life. The cell, which is equivalent to an autopoietic unit, is an organized bounded system that determines a network of reactions that in turn produces molecular components that assemble into the organized system that determines the reaction network that . . . and so on.

Varela, 1998). This characterizes the system as an autonomous identity that can be defined as auto referential. It produces its own rules of existence and therefore has a particular type of bio-logical coherence. These internal rules of the system are what define cellular life.

Criteria of autopoiesis

The most general property of an autopoietic system is the capability to generating its own components via a network process that is internal to the boundary. The boundary of the system must be "of its own making," a product of the process of component production. Whether a given system is capable of making its own boundary or not is often the most discriminating criterion by which we recognize an autopoietic system.

The question of the criteria of autopoiesis is formalized at length, but not always clearly, in the primary literature on autopoiesis. Varela, in his latest book (2000), has simplified these criteria into three basic ones, which can be expressed as follows: Verify (1) whether the system has a semi-permeable boundary that (2) is produced from within the system; and (3) that encompasses reactions that regenerate the components of the system. Thus, a virus is not an autopoietic system, as it does not produce the protein coat of its boundary or the nucleic acids (the host cell does this, and it is living). A computer virus is also not autopoietic, as it needs a computer system that is not produced by the virus itself. A growing crystal is not autopoietic, as the components are not generated from an internalized network of reactions.

Recalling the game of the two lists in Chapter 2, the baby, the fish, the tree, are all autopoietic systems. The living being is a factory that remakes itself from within. This is the common denominator of the living, regardless of whether we are looking at a micro-organism or an elephant. It must be said at this point that in its original form, the theory of autopoiesis was limited to cellular life, and Varela was for a long time opposed to generalizing it. In fact, it took quite a while before he was able to accept publication of the idea (Luisi *et al.*, 1996) that the criteria for autopoiesis could be applied to all higher forms of life, man included.

Figure 8.2 also illustrates the relationship between autopoiesis and emergence. In fact, the new properties of the bounded structure – which is life itself – arise only when the components assemble together. Life occurs only at the level of the organized, distributed ensemble. On the other hand, once these particular emergent properties of life are actualized, we have biological autonomy, one system that is capable of specifying its own rules of behavior.

The notion of "imparting its own rules" draws an equivalence between biological autonomy and auto referentiality (Varela *et al.*, 1991; Varela, 2000). In turn, auto-referentiality is related to the concept of operational closure. This is a process of circular and reflexive interdependency, whose primary effect is its own production. Operational closure must not be viewed as a lack of contact with the environment – as already stressed, any living system must be seen as an open system. The relation between autopoiesis, autonomy, and self referentiality is treated in the specialized literature, see for example Marks-Tarlow *et al.* (2001) and Weber (2002).

From all the above, it is apparent that autopoiesis belongs epistemologically to systems theory, according to which it is the organization of the components that characterizes the quality of the system. Thus, the life of a cell is a global property, and cannot be ascribed to any single component.

The living cell as an autonomous system can be seen as a "self without localization" (Varela *et al.*, 1991; Varela, 2000), as there is no single component, or single reaction, that alone is responsible for life. This concept has been found already in Chapters 5 and 6, dealing with self-organization and emergence, with regard to beehive or swarm intelligence, as well at the level of the self from the point of view of cognitive science.

What autopoiesis does not include

The molecules DNA and RNA are considered in autopoiesis only as participants in the cell's self-production and not for their ability to self-reproduce and/or evolve. Varela and Maturana often emphasize that before one can talk about the properties of life, one has to have a place to host them. The container and the logic must be there first. Likewise, in describing a car, before talking about the nature of the fuel,

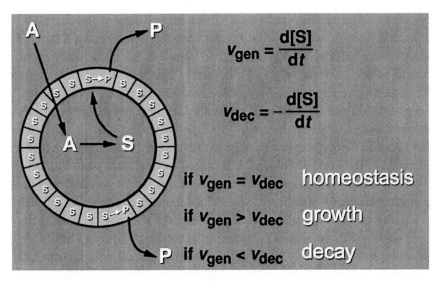

Figure 8.3 The minimal autopoietic system. This system is characterized by two competitive reactions, one that builds the components of the boundary, and another one that destroys them. According to the relative value of these two velocity constants, the system can be in homeostasis, or grow, or die.

one has to have a logical scheme describing how a car works and how motion is transmitted from the carburetor to the wheels. A classic citation by Maturana and Varela (1998) reads:

In order to reproduce something, the unit must first be constituted as a unit, with an organization that defines this unit itself. This is simple common sense logics.

Adding:

A living organism can also exist without being capable of self-reproduction.

Also, Varela asserts that to take reproduction into the definition of life would be ontologically wrong (Varela, 2000), as

reproduction is a . . .consequence of the existence of individuals. The difficult thing is to create an organism that is capable of self-reproducing with its own boundary. To divide it up in two is easy.

This notion, that reproduction is a consequence of the internal logic of life, can be visualized in Figure 8.3 (see also Luisi, 1996), which is an extension of the drawing of our Green Man (Figure 2.1) discussed in Chapter 2. It represents the various modes of existence of a minimal autopoietic system. This system is defined by a semi-permeable membrane formed by only one component S that allows the entrance of the nutrient A, which is transformed inside the system into S, the component that forms the system itself. There are two competitive reactions, one that forms S from A, with velocity v_{gen}, and one that breaks down S into the by-product P, with velocity v_{dec}. When the two velocities are numerically equal, the system is in homeostasis; when v_{gen} is greater than v_{dec}, the system can grow, and eventually give rise to a self-reproduction mode. Figure 8.3 clearly shows that self-reproduction can be seen as a particular kinetic pattern of homeostasis – a consequence of self-maintenance.

How does an autopoietic unit manage with evolution? Although evolution is not emphasized in the basic definitions, it is indeed an important part of autopoiesis. However, it is expressed in quite different terms, and, almost paradoxically, is seen as a consequence of self-maintenance: the autopoietic unit tends to maintain its own identity and has no urge to change. When the necessity to change occurs, the cell adjusts with the minimal change, so as to disturb its own identity as little as possible; and it does this by facing the external stimuli with its own internal organization. This means that only changes that are in harmony with the internal organization are accepted. If this is possible, then changes are small; if it is not possible, the organism may die. By way of this mechanism, the autopoietic unit is in constant dialogue with the environment, and over the years this builds a series of recursive couplings that form the basis of evolution.

The absence of the emphasis on DNA, self-reproduction, and evolution in the theory of autopoiesis was certainly a reason for its lukewarm reception in the community of molecular biology – a difficulty that might have been avoided, had its authors been less rigid about the matter. In fact, it is not difficult to incorporate nucleic acids and enzymes into the autopoietic scheme. This was proposed more recently (Luisi, 1993; 1997) and the corresponding modification is formally rather simple, as Figure 8.4 shows.

Finally, in the list of things that autopoiesis does not include, the term "information" should be added. This is mostly due to Francisco Varela's deep concern – which I share – about the misuse of this term in most of the current bioscience literature. When this term is not essential, it may be omitted.

Chemical autopoiesis

The term "chemical autopoiesis" indicates the experimental implementation of autopoiesis in the chemistry laboratory. The most well known of these processes is the self-reproduction of micelles and vesicles. This has been discussed in the previous chapter, where the original idea of Francisco Varela and myself was to work with bounded systems that would produce their own components due to an internal reaction, respecting the scheme illustrated in Figure 8.3. We came up with the idea of using reverse micelles (refer back to Figure 7.13) with two reagents,

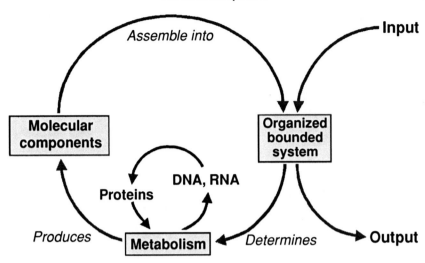

Figure 8.4 The autopoietic cycle extended to the DNA/RNA/protein world. This is the autopoietic representation of coded life.

A and B, which could react inside the boundary (but not outside) to yield as a product the very surfactant that builds the boundary (Luisi and Varela, 1990). In Chapter 7 it was also indicated how this theoretical paper led to the experimental implementation of self-reproducing reverse micelles, aqueous micelles, and vesicles (Bachmann *et al.*, 1990, 1991, 1992; Luisi, 1994; Walde *et al.*, 1994b).

All the models mentioned thus far are based on autopoietic self-reproduction experiments. The experimental implementation of a homeostatic mode of the autopoietic minimal system, which is also illustrated in Figure 8.3, proved to be much more difficult, and was realized only in 2001 (Zepik *et al.*, 2001). It is based on the oleic acid surfactant system and is schematized in Figure 8.5 (respecting the theoretical scheme of Figure 8.3): there are two competitive reactions, the reaction v_p forms oleate surfactant from the hydrolysis of the anhydride and the other reaction v_d destroys oleate via oxidation of the double bond.

When the velocities of the two reactions are numerically equal, the system is in "homeostasis," a dynamic equilibrium that does not modify the identity of the unit. Conversely, when the velocity of the building-up reaction v_p is larger than the opposite one, growth and eventually self-reproduction of the vesicles can be measured; and if instead v_p is smaller than v_d, there is destruction of the unit.

By regulating the relative concentration of the components, all three modes of reaction can be observed. The system can be made continuous by feeding nutrient oleate at the same rate at which oleate is oxidized. As far as I know, this is the only chemical system that corresponds to a homeostatic autopoietic mechanism.

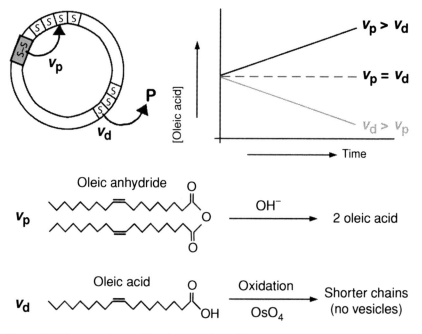

Figure 8.5 The experimental implementation of the autopoietic model of Figure 8.3 with two competitive reactions. Here one reaction forms new oleate surfactant from the hydrolysis of the anhydride and another reaction destroys oleate *via* oxidation of the double bond. Depending on whether the two velocities are equal or not, different pathways for the systems are obtained: homeostasis (which corresponds to an autopoietic self-maintenance system), growth and self-reproduction, or decay and death. (Modified from Luisi, 1993; 1996.)

This system is autopoietic, but can it be defined as living? Intuitively, one would say no. On the other hand, at first sight, this oleate system is not very different from an amoeba that absorbs a nutrient and expels a by-product. What is the difference? The question can be made more general: are all autopoietic systems living, as emphasized by Gail Fleischacker in her review on autopoiesis (Fleischaker, 1988) – or is autopoiesis just one particular subsystem?

It is important to clarify these questions; they are not entirely clear in the primary literature. To this end, the notion of cognition in the theory of autopoiesis first needs to be clarified. The question of the relation between autopoiesis and life will be discussed under "Necessary and Sufficient?"

Autopoiesis and cognition

As already mentioned, along with the question "What is life," there was another question on Maturana's agenda, namely "What is cognition?" In general, autopoiesis is concerned with organization, and cognition with the interaction with the environment. In investigating the relation between these two questions, Maturana and Varela arrived at the conclusion that the two notions, life and cognition, are indissolubly linked to each other in the sense that one cannot exist without the other.

Let us see how this works.

The starting point is the interaction between the autopoietic unit and the environment. The living unit is characterized by biological autonomy and at the same time is strictly dependent on the external medium for its survival. There appears to be an apparent contradiction here: the living must indeed operate within this contradiction.

It was said earlier that the interaction with the environment, according to the theory of autopoiesis, must be implemented on the basis of the internal logic of the living. In other words, the consequence of the interaction between an autopoietic unit and a given molecule X is not primarily dictated by the properties of the molecule X, but by the way in which this molecule is "seen" by the living organism.

As Varela puts it (Varela, 2000),

There is no particular nutrient value in sugar, except when the bacterium is crossing the sugar gradient and its metabolism utilizes the molecule so as to permit the continuity of its identity.

Actually, the compounds that the living organism extracts from the environment can be seen as something that the organism itself lacks for implementing its life. The appropriation of these missing parts is what gives "meaning" and links the autopoietic unit with its world. At this point the Santiago school introduces the term "cognition."

Varela (2000) recognizes that the choice of the term "cognition" was not an ideal one, as it has a strong anthropomorphic connotation. One thinks immediately of human cognition. According to Varela and Maturana there are, however, various levels of cognition, including those at the lower degrees of life's complexity, from unicellular to multicellular organisms, from plants to insects and fish and mammals, each with its own type and degree of cognition. Each level of cognition corresponds to a different level of life's complexity. Still, cognition is a notion that applies only to living entities, and not to the inanimate world – chemical recognition in the molecule world is not cognition.

Among all these cognitive interactions between autopoietic entities and the environment, some are particularly important because they are recursive, i.e., they happen repeatedly. For example, throughout a membrane there is a continuous flux of sodium or calcium ions. This active transport is selective in the sense that it happens with certain ions and not with others. Where does this specificity come from?. The answer lies in the phylogenesis, the history of the living species, where each state of the system at any point in time is only one moment of its history. The environment does not prescribe or determine changes to the organism, it is the structure of the living system and its previous history of perturbations that determine what reactions the new perturbation will induce.

Accordingly, changes, mutations, and evolution are seen as the result of the maintenance of the internal structure of the autopoietic organism. Since the dynamic of the environment may be erratic, the result in terms of evolution is a natural drift, determined primarily by the inner coherence and autonomy of the living organism. In this sense, Maturana and Varela's view (Maturana and Varela, 1980; 1986) is close to Kimura's (1983) theory of natural drift and to Jacob's (1982) notion of "bricolage." Evolution does not pursue any particular aim – it simply drifts. The path it chooses is not, however, completely random, but is one of many that are in harmony with the inner structure of the autopoietic unit.

These coupled interactions, accumulated over time, give a particular historical perspective to the autopoietic system. It becomes a historical product, the result of a long series of coupled interactions. As Varela says (Varela, 1989a, p. 64):

If one may consider the environment of a system as a structurally plastic system, the system and its environment must then be located in the intricate history of their structural transformations, where each one selects the trajectory of the other one.

This is not a new idea. In particular, one is reminded of Claude Bernard, who worked in the middle of the nineteenth century. He introduced the notion of "milieu interieur" (Bernard, 1865), i.e., internal milieu. The French physiologist, who is accredited with the discovery of glycogen hydrolysis in the liver, is also accredited with the introduction of the notion of homeostasis, meant as resistance to change (although the term was coined later by the American physiologist B. Cannon in the twentieth century). Noting the constancy of chemical composition and physical properties of blood and other fluids, Bernard proposed that living organisms were capable of auto-regulation, namely that they contained substances responsible for the maintenance of the internal chemical equilibrium and the function through which the organism interacts with the environment. This was the discovery of a mechanism of auto-regulation of the living organism, manifested throughout the network of reactions and exchanges with the environment (see also Lazzara, 2001). It may be noted that this principle is being rediscovered – aside from autopoiesis – over and over again: see for example the notion of Gaia (Lovelock, 1988) according to which Earth is a self-regulating (and therefore living?) organism.

In conclusion, returning to autopoiesis, each living system is a complex of circular interactions with its own environment, and this ensemble can be viewed as a continuous flow of mutual and coherent changes that have the aim of maintaining the equilibrium of self-identity. It seems clear from these considerations that for the biological cell it is the metabolism itself that is the link between the internal and the outside world. Metabolism is the result of the internal organization of the cellular system, but it is implemented by the interaction with the environment that feeds the cell and accepts its expelled by-products. Thus, metabolism itself is equivalent to cognition in the simple case of unicellular organisms. The autopoietic system and the environment change together in a congruent way, and induce changes and adaptation in each other. In this dependency "biological autonomy", can again be seen in the sense that in an autopoietic system all changes and adaptations serve to the maintenance of the structural identity. This plasticity corresponds to the very act of cognition, which is in fact based on the coupling between the autopoietic structure and the environment. For human beings, the plasticity of the brain is the motor of one particular level of cognition.

There are several other aspects related to this notion of cognition. For example, Andreas Weber discusses the relation between cognition and biological constructivism at large, and considers the semiotic aspects of this philosophy of cognition, particularly in relation to meaning (Weber, 2002, and references therein). Regarding the relation between cognition and semiotics (broadly defined as the science of signs and symbols), and from semiotic to "meaning," consider that any perturbation can be seen by an autopoietic system as a "sign" interpreted according to its inner structure.

We will come back to some of these implications later on, but I would like now to quote Weber, as he summarizes very well some of the most important concepts discussed so far (Weber, 2002):

... the Varela's school is emphasizing that the external world acts as a mere "kick", which motivates the system to establish a new equilibrium characterized only by the necessities of self-support. For a bio-semiotic approach this means that it is no longer concerned with the constraints of the mind–body-problem. Dualism becomes obsolete by the material circularity of autopoiesis. In a self-referential system, meaning is the "inner side" of the material aspect of the system's closure.

Cognition and enaction

Cognitive interaction is the result of the internal structure of the autopoietic system. In this sense, the view of Maturana and Varela is opposite to the representational model, according to which cognition is primarily the act of taking a picture of the external environment. In the autopoietic view, cognition is instead an act of mutual interaction between the inner structure of the system and the environment, in the particular sense that the environment is "created" by the sensorium of the living organism during the interaction itself.

The term "create" may occasionally have a clear physical meaning: for example, the onset of photosynthetic organisms may have indeed created a novel oxygen-rich

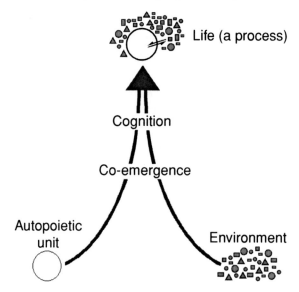

Figure 8.6 The process of enacting and co-emergence. The living structure and the environment are represented here as separate domains only for heuristic purposes. In reality, according to the theory, they are one unit.

environment. In the case of the spider, it may signify the spider's web; or the woody construction in the case of the beaver; not to mention the cities and freeways of humans. More generally, however, this term denotes the particular coupling which makes living organisms and their environment compatible and in harmony with each other. In this sense, we can accept the notion that the living structure and the environment 'create' themselves reciprocally. A qualitative rendering of this concept is given in Figure 8.6.

Here is where the thinking of Varela in particular (Varela, 1979; Varela *et al.*, 1991; Varela, 2000) comes close to certain European philosophers such as Merleau-Ponty. Consider the following statement by the French author (Merleau-Ponty, 1967):

It is the organism itself – according to the proper nature of its receptors, the thresholds of its nerve centres and the movements of the organs – which chooses the stimuli in the physical world to which it will be sensitive. The environment emerges from the world through the actualization or the being of the organism.

Not only philosophers, but modern biologists also share this view, although in their own language. For example, the well-known geneticist R. C. Lewontin (Lewontin, 1993) mentions that the atmosphere, which we all breathe, was not on Earth before living organisms and adds (p. 109):

... there is no "environment" in some independent and abstract sense. Just as there is no organism without an environment, there is no environment without an organism. Organisms do not experience environments. They create them. They construct their own environments out of the bits and pieces of the physical and biological world, and they do so by their own activities.

Furthermore (p. 63):

A living organism at any moment in its life is the unique consequence of a development history that results from the interaction of and determination by internal and external forces. The external forces, what we usually think as "environment", are themselves partly a consequence of the activities of the organism itself as it produces and consumes the conditions of its own existence.

Varela coins the word "enaction" (Varela, 1979; Varela *et al.*, 1991; Varela, 2000) to indicate this very process of mutual calling into existence: the organism with its *sensorium* "creates" its own world; the environment allows the living organism to come into being. Thus the term enaction indicates a mutual process of adaptation as a co-emergent act. Actually, the term "co-emergence" is also appropriate and can be used as an alternative to enaction. Again: the emphasis and the overall concern of the enactive approach is not to define cognition in terms of an objective relation between a perceiver and a world, but rather to explain cognition and perception in terms of the internal structure of the organism. It is also relevant to repeat that cognition is defined at various hierarchic levels: cognition in micro-organisms, in multicellular organisms, fish, mammals . . . , each time acquiring a higher level of complexity and sophistication. At the level of mankind, cognition may become perception and consciousness, but the same basic mechanism is operative at all of these hierarchic levels of life. In the process of enaction, the organic living structure and the mechanism of cognition are two faces of the same phenomenon, "the phenomenon of life" (Varela, 2000).

Finally, it should be mentioned that the interaction between organisms and their environment may be part of the more general scenario of ecology. If we accept that living organisms make and continuously change the environment in which they live, and vice versa, we must also accept the idea that the world is constantly changing and cannot exist without changing. (Lewontin, 1993; Capra, 2002).

Necessary and sufficient?

We go back now to the question of whether and to what extent autopoiesis is the necessary and sufficient condition for cellular life. In the early days of autopoiesis, Maturana and Varela explicitly wrote (Maturana and Varela, 1980, p. 82) that: "autopoiesis is necessary and sufficient to characterize the organization of living systems" and Gail Fleischaker, in the previously cited review on autopoiesis

(Fleischacker, 1988), writes that whatever is living must be autopoietic, and that conversely, whatever is autopoietic must be living.

As a matter of fact, accepting the two assertions in the primary literature, one that autopoiesis is sufficient to characterize the organization of life, and the other, that cognition is equivalent to life, the conclusion should be reached that each autopoietic system is cognitive, and therefore living. This is actually what Fleischacker refers to.

I believe, however, that the assertion that every autopoietic system is living goes too far and in this section I would like to clarify the limits of this assertion. In doing so, I follow the lines of a paper recently presented in collaboration with Michel Bitbol (Bitbol and Luisi, 2004). The question of whether autopoiesis is the necessary and sufficient condition, or only the necessary one, has also been asked by Bourgine and Stewart, (2004); and earlier, in another context, by Weber (2002).

To clarify this point, it may be useful to keep in mind two practical cases, which have been mentioned previously: a synthetic vesicle that absorbs a particular molecule from the medium, and by so doing, reproduces itself via an autocatalytic process; and – second case – a bacterium that "recognizes" and absorbs sugar from the environment. At first sight, these two processes are similar; however, we would commonly ascribe the definition of living to the bacterium, and generally not to the other case. Intuitively, the difference can be seen in terms of cognition. How can we clarify this point?

We have said that cognition is definitely *not* a passive picture of an external reality. It is instead mostly governed by the internal organization of the cognitive system and in particular is an act of mutual co-participation between environment and organic (cognitive) structure (see also Bitbol, 2001).

Also the point has been made earlier that cognition is linked to metabolism. In particular, metabolism corresponds to a dynamic interaction with the outer medium and therefore is the biological correlate of cognition. Very important, and this is now fundamental for our present issue, is a pre-existing metabolic network (which can also be extremely simple) that *recognizes* the external molecule from the environment and makes it part of its being. Generally, the molecule that is being selected is an already well-known metabolite, such as the sugar for the bacterium. This is actually the classic, simpler form of metabolism, although more complex levels can be envisaged.[1]

With this observation in mind, let us consider again our two examples, the bacterium and the autopoietic vesicles (see Figure 7.16), again asking the question of

---

[1] In addition to this "known" aspect of metabolism, there is another one, albeit less frequent, that refers to the interaction of *novel* compounds. This is important, as it gives the possibility of permanent changes, and is associated with adaptation and evolution. In other words, one should consider two aspects of metabolism/cognition: the normal homeostatic metabolism, which corresponds to the normal life and self-maintenance of the cell from within; and a metabolism of "novel" elements that may operate changes in the structure.

whether the notion of cognition so qualified can permit a discrimination between the two systems.

Is the vesicle system a cognitive system? The answer is negative. In fact, there is no pre-existing "metabolism" that selects out the binding and the interaction with S–S; S–S is not recognized by a reaction system that is already in the structure, S–S is not accepted and assimilated as a part of an existing "metabolic network." Even when the next steps of the hydrolysis reaction take place on the bilayer, the further addition of S–S does not proceed because of a pre-existing reaction network – but because of simple hydrophobic forces, which operate regardless of the internal reactivity of the vesicle. The vesicles do not even reach the first stage of cognition.[2] In these terms, the self-replicating micelles and vesicles are not cognitive systems although they are simple autopoietic systems.

Of course, it is apparent that this conclusion is based on a particular definition of metabolism/cognition, albeit quite general and simple. However, I believe that this discrimination is useful and in fact it can help to clarify the difference between the feeding bacterium and the feeding vesicle.

From the above, one can elicit that autopoiesis is not a necessary and sufficient condition for life. It is a necessary condition, but then it takes cognition, at least in the simplest stage, to arrive at the process of life. The union of autopoiesis and the most elementary form of cognition is the minimum that is needed for life.

As already mentioned, Bourgine and Stewart (2004) arrive at the same conclusion, based on an elaborate and elegant mathematical treatment. They state in fact that autopoiesis is not a necessary and sufficient condition for life. For the finer differences between these two treatments, see Bitbol and Luisi (2004), where broader implications of the definition of cognition are also discussed.

Having said that an autopoietic self-reproducing vesicular system is not a living system, the question may arise as to whether and at which point such a vesicle system may reach the stage of cognition, and therefore be called living. This is an important point, because it may suggest experiments in wet biochemistry to implement minimal autopoietic and cognitive systems.

To this aim, let us consider Figure 8.7, keeping in mind also footnote 1. There is an internal cycle of three components, A, B, and C, and all this forms an autopoietic unit, in the sense illustrated in Figure 8.3, whereby, for example, the substance C is the membrane component. Let us now consider a substance X–Y that interacts with the autopoietic unit and is not recognized by the metabolic cycle. This

---

[2] This point is considered in more detail by Bitbol and Luisi (2004), where a correspondence between these two stages of metabolic cognition and the general two-step cognition scheme developed by Piaget (1967) is also illustrated. The terms assimilation and adaptation are used in this context to indicate the two stages of cognition. The main difference of course is that Piaget started from complex human cognition as a model for biologically more elementary forms of cognition, whereas Varela and Maturana proceeded the other way around. I thank Francesca Ferri of the University of Bologna for discussion of these concepts.

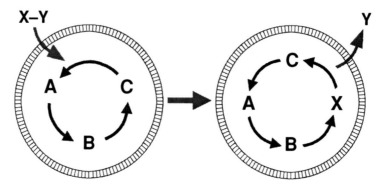

Figure 8.7 The foreign substance X–Y enters in the autopoietic structure and is capable of being incorporated in the existing metabolism, modifying it (of course via and due to the corresponding genetic changes). This corresponds to adaptation and to a permanent modification that may give rise to evolution.

molecule can be absorbed and parked inside the unit without being integrated, and eventually be expelled. It can also block one of the reactions of the cycle (i.e., act as an inhibitor or a poison). Or else, as shown in the figure, it can become part of the metabolic cycle. This corresponds to a change of the genomic system, e.g., in a cell this would imply the development of novel enzymes that are capable of accepting and transforming the molecule X–Y and inserting it in the cell metabolism.

As a consequence, and referring again to Figure 8.3, the overall values of the constant v_{gen} and/or v_{dec} can be changed. This corresponds to a change in structure but not of autopoietic organization. It corresponds also to the view of Bourgine and Stewart (2004), according to which there is cognition when there is a cause and a resulting effect.

This visualization of Figure 8.7 is pragmatically important: as already mentioned, it may suggest to the experimentalist some minimal cell that can be fabricated in the laboratory. In fact, if we could realize a vesicular system hosting in its interior the "simple" metabolic cycle of Figure 8.7, then we would have realized a minimal cognitive system, which is autopoietic – and therefore we would reach the conclusion that such a system would, indeed, be living. Of course this kind of synthetic life would be the first step of a graduation of complexity. In the same way that the term cognition has a graduation of complexity, the term life obviously has various levels of complexity.

One glance further up: from autopoiesis to the cognitive domain

The notion of cognition, as one of the determinants of life, permits a link with the humanistic cognitive domain, including ethics and consciousness. I have mentioned this in the preface, and I would like to express this concept briefly here. These ideas derive from the work of Maturana and Varela, and more specifically from Varela's books (Varela, 2000; Varela *et al.*, 1992; see also Varela, 1989).

The starting point is the consideration that cognition, as defined above, is a stratified concept – namely there are different hierarchic levels of cognition depending upon the complexity of the living organism. There is cognition at the level of the amoeba, then at the level of a bee, of a dog, a chimpanzee, and at the level of mankind. As the sensorium develops, and flagella and legs and eyes and fingers and brain develop, correspondingly various and more differentiated levels of cognition emerge. Eventually they may acquire the form of perception, intelligence, or consciousness (for the moment meant simply as a mental activity, and not as self-awareness). Here, at the level of humans, the main point is again that cognition is not based on the representation of external objects that our mind is visualizing. Rather, there is a recursive coupling in operation, according to which the experience of the outside world is possible only because of the nature of the human structure. It is human consciousness that makes possible the emergence of objects of the outside world.

The city in which we live, the value ascribed to plants and fruits, to love and politics and religiosity, are all products of our mental cognition – also the acquisition of indirect evidences (a photo, a third-person report) is based on our consciousness. Conversely, consciousness is created by the world we live in; more specifically, by the experience and learning that we have in our lives. Going back to the example of whether a rose is only molecules or not – it is clear that a rose takes its essence from the concept of "rose" that is present in our consciousness. This is the result of the experience of vision, smell, poetry, musical tradition, culture – a rose makes no sense without human consciousness. For a fish, a rose is not a rose.

Thus, a mechanism similar to that illustrated in Figure 8.6 can be proposed. This is shown in Figure 8.8, giving a pictorial representation of the notion of embodied mind. The term "embodied mind" has become popular in the cognitive sciences, meaning that the notion of mind makes sense only when it is realized in a physical embodiment – and vice versa: that our bodily experience and notion of life is the product of consciousness. All this, as already mentioned, is well accepted in most of the cognitive sciences today. It is just a pity that the work of Varela, again, is not duly cited in this respect in the references of main stream cognitive science. Varela's approach is based on phenomenology, a philosophical concept that goes back to Husserl, and in fact Varela and coauthors (Varela *et al.*, 1991) devote a large section to this philosopher in their books, *The Embodied Mind*. They also make clear (Maturana and Varela, 1998; Varela, 2000; Thompson and Varela, 2001b) that this view of the reality makes the dualistic view of the Cartesian world obsolete, as there is no body without mind, and vice versa.

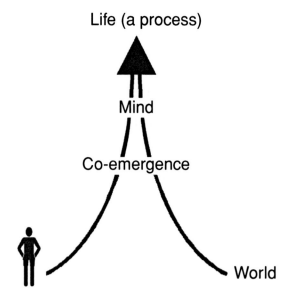

Figure 8.8 The co-emergence of the human autopoietic structure and the environment in the process of life. This co-emergence (cognition) in this case becomes the
"embodied mind" (Varela *et al.*, 1991) – as mind (and eventually consciousness) –
is based on experiencing the world. Body and mind are two faces of the same phenomenon, the phenomenon of life (see Varela, 2000). However, this schematization
assumes a starting duality that is not supposed to exist in reality.

Figures 8.6 and 8.8 are both representations that suffer from the limits of our
informational technique, as they show two distinct things at the bottom, implying an
initial separation between living organisms and their environment. This separation
is just what the notion of co-emergence negates. The complementarity between
yin and yang in classic Chinese philosophy comes to mind: they also cannot be
separated from each other; or the complementarity between wave and particle in
quantum mechanics; they can be distinguished from each other only when carrying
out a specific experiment.

We can go yet another step further in this cautious tapping at the higher domains.
One point is linked to the trivial observation made earlier, that a rose is not a rose for
a fish. In fact, this is tantamount to saying that there are as many realities as there are
different observers, since each of them will have a different internal structure, and
therefore a different co-emergence with the environment. When extrapolated, this
takes us to Varela's notion of "groundless-ness," the lack of an objective, unique
reality; and to the even more complex concept of the simultaneous existence of
many different worlds.

Some readers may recognize in all these ideas a flavor of Buddhism. I mention this
because the life and thoughts of Francisco Varela were indeed significantly
influenced by Buddhism. *The Embodied Mind* (Varela *et al.*, 1991) is in fact a book

which, among other things, proposes an ambitious synthesis between Buddhism and cognitive science.

Aside from that, a general corollary to what has been asserted thus far, is that consciousness is not a transcendental property coming from some higher place, but is instead a property inborn in the essence of the being, something that has its origin and roots in the physical body. More specifically: if we accept that life comes from inanimate matter, and if we accept the Darwinian view according to which the higher forms of life come from the evolution of simpler ones, then also consciousness comes "from within." In other words, it starts from the very autopoietic organization of the living and can be seen as an emergent property of the more complex forms of life. This emergence is not, however, a mechanical outlet of the brain complexity (the more traditional view) but it is the result of the co-emergence (enacting) illustrated above.

If consciousness comes 'from within', then, of course, the same can be said about all "by-products" of consciousness, such as ethics, spirituality, and the very notion of God. They all would be, in this perspective, the noble products of consciousness. Hardly a novel idea, but what is new in this case is that such a derivation comes straight from the biology of autopoiesis. It is then a form of "lay spirituality," in the sense that it is not based on a mystical delivery from above but comes from the same origin as the origin of physical life.

To some extent, these ideas about consciousness are present in modern cognitive science, although with different forms and terminology, and the interested reader can refer to more specialized literature, for example the work by Damasio (1999) or by le Doux (2002). There are now many books on the subject of consciousness and many novel academic institutions devoted to the study of consciousness, with much emphasis on the relation between brain and mind. This is certainly remarkable in an area dominated by the molecular paradigm. Very little has yet been done to connect this with a bio-logical theory of life as a property from within, but I believe that the trend will move in this direction. In this sense, Francisco Varela has again been somewhat of a pioneer.

Social autopoiesis

We go back now to basic autopoiesis, in order to consider what came as an interesting, and unexpected, development of this theory. By "unexpected," I mean that even the authors of the Santiago school had not foreseen it. This is "social autopoiesis."

The main feature of autopoiesis is self-maintenance due to a process of self-generation from within. Although this concept came from the analysis of a living cell, it can be metaphorically applied to social systems.

Consider, for example, a political party, or a family, whereby the rules that define a party or a family can be seen as a kind of boundary given by the social structure

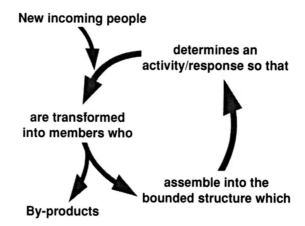

New incoming people

determines an activity/response so that

are transformed into members who

assemble into the bounded structure which

By-products

Figure 8.9 Towards social autopoiesis: the cyclic logic of autopoiesis applied to social systems. Notice the analogy with Figure 8.2. The transfer from biology to social science assumes that "human relationships" substitute for the chemical interactions among the cell constituents and that the definition rules of the social community substitute for the membrane boundary.

itself. The whole system enjoys a dynamic equilibrium, as certain members leave the structure, and new people come in, and are transformed into steady members by the binding rules of the party or of the family. There is a regeneration from within, there is the defense of the self-identity; the metaphor of the living cell applies. Also, in all these systems, certain characteristic features of biology can be recognized, such as the notion of emergence – the family being an emergent property arising from the organization of single individuals, etc. It is also apparent how the notion of autonomy can be applied to such social systems. In fact, they are characterized by their own internal laws, which are valid within their boundaries.

Figure 8.9 gives a qualitative picture of this situation, analogous to the illustration seen earlier (Figure 8.2) about the life of a cell.

When the social sciences picked up on this idea they stirred up a great deal of intellectual excitement. The German sociologist N. Luhmann constructed an entire field based on social autopoiesis (Luhmann, 1984), and autopoiesis was also applied to the judicial system, to literature, and in systemic family therapy – see the work by John Mingers (Mingers, 1992; 1995; and 1997).

Varela remained somewhat skeptical about these extensions of autopoiesis. He says in this regard (Varela, 2000):

These ideas are based, in my opinion, on an abuse of language. In autopoiesis, the notion of boundary has a more or less precise meaning. When, however, the net of processes is transformed into one 'interaction among people', and the cellular membrane is transformed into the limit of a human group, one falls into an abuse, as I expressly said.

Recently Maturana has expressed similar thoughts (see Poerksen, 2004). Despite the authors' doubts, the fact that something grows out of their ideas and produces new viewpoints and fields of inquiry is all for the best.

Autopoiesis and the chemoton: a comparison of the views of Ganti with those of Maturana and Varela

One advantage of the idea of autopoiesis is its extreme simplicity. This comes at the cost of a lack of structural and mechanistic details. Ganti's chemoton provides a more detailed and more complex view of the unit of cellular life (Ganti, 1975; 2003). Let's now sketch the basic theory of the chemoton in order to draw a comparison with autopoiesis.

Interestingly enough, the papers on autopoiesis and the chemoton both appeared in 1975, and they share a similar history, in that both had difficulty in gaining acknowledgment in the mainstream of international biochemistry. In Tibor Ganti's case, one reason was his isolation in Hungary, which at that time did not readily permit contact with the scientific world at large. In addition, diffusion of the chemoton ideas was delayed for the same reasons mentioned for autopoiesis, most notably the fact that they are both systemic views where nucleic acids are not presented as the main heroes.

The chemoton scheme is illustrated in Figure 8.10. It consists of three subsystems: a metabolic autocatalytic network, a bilayer membrane, and a replicable information carrier molecule, or template. Here, as in autopoiesis, the membrane is an important player. However, what happens within the membrane is depicted in more detail. In particular, looking at Figure 8.10, note that the entire system is fed by nutrient X which first feeds the metabolic network with the various A_i molecules. This is an autocatalytic chemical cycle, autocatalytic because two A_1 molecules are produced from the original single A_1 molecule.

From this first metabolic cycle, two products are formed, which feed the other two cycles: T^1 is the precursor of the membranogenic molecule T, and the association of several T molecules gives rise to the membrane self-assembly depicted as T_m. This membrane self-assembly can grow and divide spontaneously. From the A cycle stems the product V^1 as well, which makes a polymer pV_n of n molecules of V^1. The polymer pV_n undergoes template replication; R is a condensation by-product of this replication and it is needed to transform T^1 into T.

At first sight, the system may appear rather cumbersome and complicated; Ganti's claim (Ganti, 1975; 2003) is that this scheme represents the simplest rendering of a minimal living chemical system. In fact, the model is arranged such that the various components interlock with each other: the growth of the membrane occurs only if the template is replicated, and the membranogenic T is formed from T^1 only in

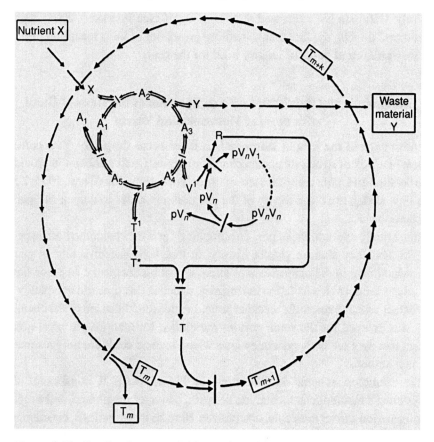

Figure 8.10 Ganti's chemoton (redrawn from Maynard-Smith and Szathmáry, 1995, based on the original by Ganti, 1984). The metabolic subsystem, with intermediates $A_1 \rightarrow A_2 \rightarrow \ldots \rightarrow A_5$, is an autocatalytic chemical cycle, consuming X as nutrient and producing Y as waste material; pV_n is a polymer of n molecules of V^1, which undergoes template replication; R is a condensation by-product of this replication, needed to turn T^1 into T, the membranogenic molecule; the symbol T_m represents a bilayer membrane composed of m units made of T molecules.

the presence of R, which is a by-product of replication. In other words, there is a coupling between replication and membranogenesis.

Maynard-Smith and Szathmáry (1995) discuss at length some of the implications of the chemoton, for example, the question of how the chemoton has heredity. They point out some of the difficulties in implementing the chemoton but conclude by saying (Maynard Smith and Szathmáry, 1995):

that this abstract system . . . is an excellent mental jumping-board for understanding the origin of life.

This points to another difference between autopoiesis and the chemoton: that autopoiesis was not born to tackle the problem of the origin of life, but to describe what life is. On the other hand, the great simplicity of autopoiesis has allowed the implementation in the chemical lab of autopoietic systems; its generality has permitted the rise of social autopoiesis; and the philosophical tissue in which it is embedded has permitted the link with the concepts of emergence, biological autonomy, referentiality, up to the notion of cognition, and even a bridge to cognitive science. Autopoiesis and the chemoton are two different and brilliant ideas, with different potentialities; however, they were born out of the same "Zeitgeist," to look at cellular life from the point of view of system biology.

Concluding remarks

The theory of autopoiesis is based on the observation of the actual behavior of a living cell. As such, it is not an abstract theoretical model for life – there are many of these – but a deductive analysis of life as it is on Earth. It is in a way a picture of the blue-print of cellular life, and it is fascinating to see how many concepts related to the process of life – emergence, homeostasis, biological autonomy, interactions with the environment, cognition, evolutionary drift, etc. – pour forth from this analysis in a coherent way.

The main ingredient of this unity is the fact that all is seen "from within," i.e., from the logic of the internal organization of the living system. As soon as the autopoietic unit reaches the complexity of biological autonomy, everything that happens within the boundary, as well as the perturbing events from the outside, are interpreted and elaborated in order to maintain the identity of the living.

The other important basic notion is the "enacting," the process according to which there is a mutual co-emergence of autopoietic structure and environment. One triggers the essence of the other in a complex process of interaction. We have also seen that the notion of cognition permits a bridge between the biology of cellular life and the cognitive sciences. I maintain that autopoiesis is the only available simple theory that is capable of providing a unified view of life from the molecular level up to the level of human perception.

Despite the richness of this envelope, autopoiesis does not have a large impact on mainstream biological science. Why is this so? This question has been partly answered already: autopoiesis originated in a time-window (the early 1970s) when the world of biology was completely dominated by a vision of DNA and RNA as the holy grail of life. Alternative views about the mechanism of life didn't have much chance of appearing in mainstream journals. This argument also holds for Ganti's chemoton theory. In the case of autopoiesis, this situation also reflects the intellectual choice of the authors, and partly perhaps their rigidity, as it would have taken relatively little to make autopoiesis more harmonious with the DNA/RNA world.

There are some signs in the modern life sciences literature to indicate a return to a system theory of life processes, which emphasize collective, integrating properties – such as self-organization and emergence. In this new – perhaps more philosophically – mature Zeitgeist, autopoiesis could re-emerge as a very useful conceptual framework.

Questions for the reader

1. Do you know of any living system that does not obey the criteria of autopoiesis?
2. Do you accept the idea that self-maintenance is the most basic property of cellular life, self-reproduction being a particular kinetic outcome of self-maintenance?
3. Do you find autopoiesis satisfactory as a blue-print of cellular life? (If not, what do you feel is missing?)
4. Do you accept the notion that cognition is a stratified property with different hierarchic levels of complexity – going from micro-organisms to human beings – to become here perception and consciousness?

9

Compartments

Introduction

The following chapters are rather technical, in contrast to most of the previous ones, which have been more discursive. As already mentioned, this reflects the dual nature of the field of the origin of life, which is based both on the "software" of epistemological concepts, and on the hardware of organic and physical chemistry.

We are approaching the final part of the book, concerned with cellular models based on vesicles. The main keywords are now "compartment" and (if this word exists) "compartmentation." The biological potential of these aggregates is closely related to their physical properties, and for this reason some of these basic characteristics will first be briefly considered. Also, to give a proper background to these properties, it may be useful to compare various kinds of compartments, such as micelles, reverse micelles, cubic phases, and vesicles. This will be useful to understand better biochemical reactions in vesicles, which will be dealt with in the next chapter.

Surfactant aggregates

We have already seen in Chapter 5, on self-organization, how and why amphiphilic molecules tend to form aggregates such as micelles, vesicles, and other organized structures.

Figure 9.1 illustrates again this general phenomenon, emphasizing that for a given surfactant the type of aggregate formed is determined by the corresponding phase diagram, namely by the relative concentration of the three (in this case) components.

The molecular structure of the surfactant influences the form of the aggregate, and there are some geometrical empirical rules (Israelachvili *et al.*, 1977, Israelachvili, 1992) illustrated in Figure 9.2, based on the geometrical parameters of the surfactant molecule. In particular the volume V occupied by the surfactant, the head area

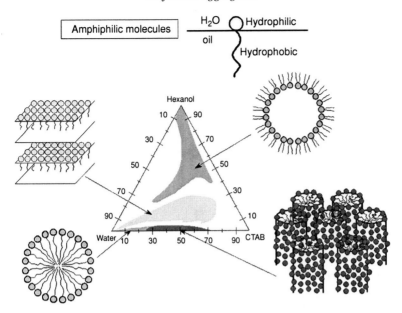

Figure 9.1 The phase diagram for the positively charged surfactant CTAB (cetyltrimethylammonium bromide): depending on the relative concentration of CTAB, water, and hexanol, quite different organized structures are formed.

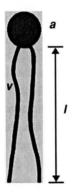

Figure 9.2 Calculation of the surface packing parameter, $V/a \times l$, which determines to some extent the form of the aggregate. For $V/(a \times l)$ *c.* 1, bilayers are preferentially formed; when this ratio is less or more than one, spherical aqueous and reverse micelles are formed, respectively. Self-assembly may be described in terms of the curvature that exists at the hydrocarbon–water interface.

a, and the length of the tails, *l*, are considered. Accordingly, when the ratio $V/(a \times l)$ is around unity, bilayers are formed; when this ratio is less than one, tendentiously spherical aqueous micelles are formed, and when greater than one, reverse micelles are preferentially formed. In reality, however, it is not always possible to predict the form of aggregate on the basis of these considerations, and the relation

Table 9.1. *Some different types of surfactants with their basic properties*

Surfactant	cac (mM)[a]	N_x^b
Anionic		
Sodium dodecylsulfate	8.1	62
Sodium dodecanoate	24	56
Sodium cholate	14	3
Cationic		
Hexadecyltrimethylammonium bromide (CTAB)	0.92	61
Dodecylammoniumchloride	15	55
Non-ionic		
Octyl glucoside	25	84
Dodecyl(polyethylenglycol(23)-ether (Brij 35)	0.09	40
4-ter-octylphenyl(polyethylenglycol(9/10)-ether (Triton X-100)	0.31	143
Zwitterionic		
Sodium taurocholate	3.1	4
N-dodecyl-N,N-dimethylammonium-3-propansulfonate (Sulfobetain SB12)	3.3	55

[a] cac: critical aggregate concentration, above which aggregates (micelles and vesicles) are formed.
[b] N_x: average aggregation number.

between molecular geometry and thermodynamics continues to be rather elusive; also because a series of environmental factors (salt, pH, and temperature) affect the form of aggregate.

Depending upon the nature of the polar head group, surfactants are usually classified as anionic, cationic, non-ionic. Table 9.1 shows a few examples with the corresponding cac (critical aggregate concentration) and the aggregation number, N_x (Pfüller, 1986; Carey and Small, 1972). Surfactants have a very large technical importance, for example in laundry, cosmetics, the food industry, oil refining, industrial cleaning and processing (Falbe, 1987). The USA production in 1990 alone amounted to *c.* 4 million tons per year. Roughly two thirds of the market is taken by anionic surfactants, the rest mostly by non-ionic surfactants; a small share is taken by cationic surfactants.

The basic common denominator for all these applications is qualitatively well understood: surfactants and their aggregates permit mixing, or at least close interaction, between phases or substances that are per se immiscible with each other – mostly oil and water. This is how grease is washed off from our hands when we use soap, the removal being mediated by micelles. In turn, micelles and vesicles permit the formation of an extraordinarily efficient interfacial system. Figure 9.3 gives a dramatic demonstration of this, showing that the total surface of a concentrated soap solution in your sink may well correspond to the surface of a stadium!

Micellar surface
$A = 19.6 \times 10^{-14}$ cm^2

$r = 12.5$ Å

Spherical micelle of caprylate ions

[Micelles](mol l^{-1})	1.7×10^{-10}	8.5×10^{-8}	8.5×10^{-5}	1.0×10^{-3}
Micellar surface in a litre of solution	19.6 cm^2	1.0 m^2	10^3 m^2	0.012 km^2
Equals a surface of a	Passport photo	Desk	Swimming pool	Stadium

Figure 9.3 Calculation of the interfacial area formed by an anionic surfactant (caprylate) giving rise to micelles.

It is no surprise then that surfactants are used so extensively in technical applications; and of course this large surface area achieved by surfactant aggregates immediately inspires ideas of applications in basic chemistry too: if for example the surfactant head had some catalytic properties, these could be extended and developed into an almost incredible dimension.

As far as chemistry and life sciences are concerned, there are for me and many others two main reasons for this fascination, summarized in Figure 9.4: firstly, above a certain critical concentration, structural order is achieved starting from the chaotic mixture of disordered surfactant molecules. As discussed earlier, this increase of order is attended by an increase of entropy and a decrease of free energy.

Secondly, there is the emergence of compartments, with an inside that is physically different from the outside. The discrimination between inside and outside, applicable to compartments, is the first structural prerequisite for the living cell and the living in general. All these notions are beautifully applied to nature. Life is dominated by water as a background medium; however, living beings contain a considerable amount of lipophilic compounds that are per se insoluble in water – lipids in particular. The compatibility of lipids with water is made possible by the concomitance of two almost opposite effects: a process of de-mixing, by which lipids make their own microphases avoiding contact with water as much as possible; and the exposure of the organized hydrophilic head groups to water, by virtue of which the microphase becomes soluble – or at least compatible with water. Figure 9.5 shows the well-known structural analogy between the biological membrane and liposomes. This principle of separating hydrophobic microphases and rendering them water compatible by exposing hydrophilic moieties to water can be seen in other structural biological domains, for example in proteins (formation of

(a) **Self-organization**

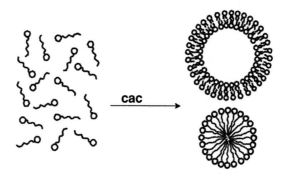

cac

(b) **Compartmentation**

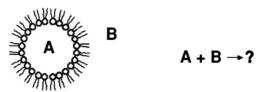

A

B

A + B →?

Figure 9.4 Two good reasons for fascination in the field of surfactants: (a) spontaneous order out of a chaotic mixture of monomers; and (b) the emergence of compartments (microheterogeneous reactions . . .). (Vesicle and micelle are not to scale.)

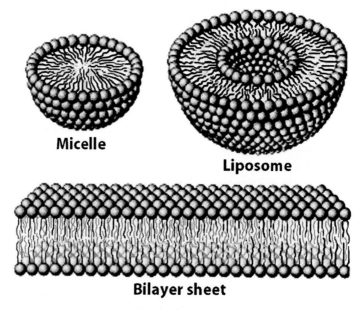

Micelle

Liposome

Bilayer sheet

Figure 9.5 Views of the structures that can be formed by phospholipids in aqueous solution. The grey spheres depict the hydrophilic heads of phospholipids, and the squiggly lines the hydrophobic tails.

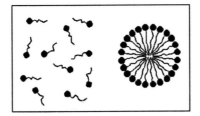

Figure 9.6 Aqueous micelles from sodium dodecylsulfate (SDS) and their physical properties. Average radius of a micelle (R_H), 2.2 nm; average aggregation number, 62; approximate relative mass of a micelle (M_r), 1.8×10^4; average half-life of a SDS molecule in the micelle, 0.1 ms; CMC (25 °C, H_2O), 8.1×10^{-3} M; i.e., monomer concentration by 10 g SDS l^{-1} (35 mM), 2.3 g l^{-1}.

hydrophobic domains or patches) and with the aromatic bases in DNA (segregated in the interior core of the double helix, avoiding water).

Let us now see some basic features of the various compartments.

Aqueous micelles

Aqueous micelles have diameters ranging typically from 0.5 to 5 nm, and being so small, do not scatter visible light and form transparent solutions. Figure 9.6 shows some basic parameters for aqueous micelles, relative to the well-known SDS (sodium dodecylsulfate). Micelles are thermodynamically stable, and this is a significant difference with respect to most large vesicle aggregates.

The thermodynamics of micelle formation has been studied extensively. There is for example a mass action model (Wenneström and Lindman, 1979) that assumes that micelles can be described by an aggregate M_m with a single aggregation number m, so that the only descriptive equation is: $mM_1 \leftrightarrows M_m$. A more complex form assumes the multiple equilibrium model, allowing aggregates of different sizes to be in equilibrium with each other (Tanford, 1978; Wenneström and Lindman, 1979; Israelachvili, 1992).

In the mass action model the micellar system can be described by only one parameter, and despite this simplicity, a good qualitative description of the main physical properties is obtained, for example the onset of cmc (critical micelle concentration), as shown in Figure 9.7. Notice that the formation of micelles becomes appreciable only at the cmc, and after that, by increasing further the surfactant concentration, all added surfactant is transformed directly into micelles, so that the surfactant concentration in solution remains constant at the level of cmc.

Hydrophobic substances added to an aqueous micellar solution tend to be entrapped into the "oily" interior, and if the substance contains a fluorofor, its fluorescence properties may change drastically upon entrapment, due to the change

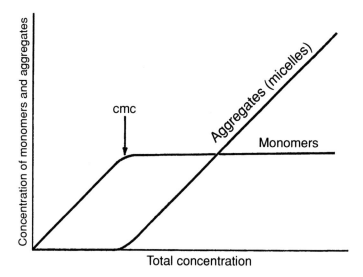

Figure 9.7 Concentrations of monomers and micelles as a function of total concentration (schematic). Most single-chained surfactants containing 12–16 carbons per chain have their cmc in the range $10^{-2}-10^{-5}$ M, while the corresponding double-chained surfactants have much lower cmc values due to their greater hydrophobicity. (Adapted from Tanford, 1978). Some important cmc values are listed, as cac values, in Table 9.1.

in the environment. Figure 9.8 gives an old and classic example of this phenomenon and shows how this can be used to determine the cmc of sodium laureate micelles. This compartmentation of added substances can also take place on the polar surface of the micelles, and this effect, or the combination of the two types of forced compartmentation, can give rise to the so-called micellar catalysis. There are examples of this in Chapter 7, on self-reproduction, in particular in the case of the self-reproduction of aqueous micelles. This mechanism is due to water-insoluble surfactant precursors being bound to the membrane of the micelles, and hydrolyzed there by micellar catalysis.

Micellar catalysis is a broad field (Fendler and Fendler, 1975; Rathman, 1996; Rispens and Engberts, 2001), and caution is needed when using this term. In fact, whereas the broad term "catalysis" is justified when referring to an increase of the velocity of reaction, this does not always mean that the velocity constant is increased (namely that there is a decrease of the specific activation energy). Rather, the velocity effect can be due to a concentration effect operated by the surface of the micelles. This is also the case for the autocatalytic self-reproduction of micelles discussed in the previous chapter, where the lipophilic precursor of the surfactant is concentrated on the hydrophobic surface of the fatty acid micelles (Bachmann *et al.*, 1992), a feature that has given rise to some controversy (Mavelli and Luisi, 1996; Buhse *et al.*, 1997; 1998; Mavelli, 2004).

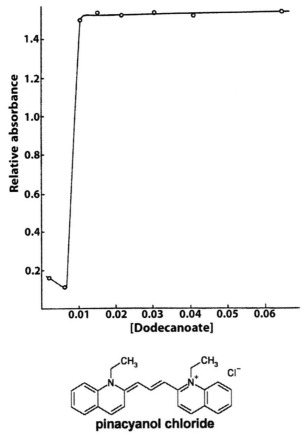

pinacyanol chloride

Figure 9.8 The absorbance of 1.05×10^{-8} M pinacyanol chloride at 610.0 m in pH 9.59 sodium borate buffer (I = 0.1) at 50 °C vs. dodecanoate concentration. The absorption spectrum of pinacyanol chloride in aqueous solution of anionic soaps changes sharply to one characteristic of its solutions in organic solvents within a small range of soap concentration ($\lambda_{max} \sim 610$ nm). This effect is attributed to the formation of micelles, in whose hydrocarbon-like layers or cores the dye is solubilized. The concentration of soap at which this spectral change occurs is taken as the cmc. The use of dyes for the determination of cmc values may lead to micelle formation at a concentration below the "true" cmc. In practice, the method gives only a rough approximation of the cmc. (Adapted, with some modifications, from Corrin *et al.*, 1946.)

Compartmentation in reverse micelles

Reverse micelles form in aprotic organic solvents, such as hydrocarbons or CCl_4, and can be seen as a core containing water (the water pool) solubilized in an oily environment (for example hydrocarbons) by the hydrophobic tails. Figure 9.9 also shows the structure of AOT (from aerosol octyl), which is the most popular surfactant for reverse micelles. A typical reverse micellar system appears as a clear

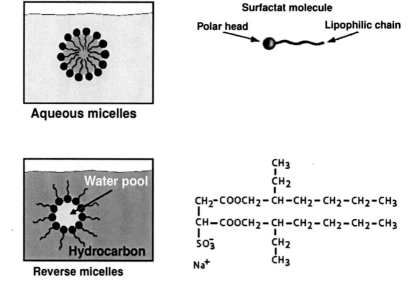

Figure 9.9 Schematic representation of aqueous and reverse micelles (cross sections), with the structure of the most popular surfactant for reverse micelles, the AOT i.e., bis(2-ethylhexyl)sodium sulfosuccinate. The typical conditions to obtain reverse micelles are as follows: isooctane, 25–1000 mM AOT, 0.2–2% water, $W_0 = [H_2O]/[AOT]$.

hydrocarbon solution, and the water content can usually be varied within an order of magnitude, say between 0.5% to 5%. The amount of water that can be solubilized is generally expressed as molar ratio w_0 between water and surfactant molarities. This ratio determines the size of the water pool and the stability of the micelles in a given solvent (for this and other properties of reverse micelles see Pileni, 1981; Boicelli *et al.*, 1982; El Seoud, 1984; Luisi and Straub, 1984; Luisi and Magid, 1986; Levashov *et al.*, 1989).

Micelles are thermodynamically stable; at the same time they are highly dynamic systems, as they break and re-form rapidly, and exchange material by continuous collisions and fusion processes (Fletcher and Robinson, 1981; Harada and Schelly, 1982). Figure 9.10 gives an illustration of this for reverse micelles, but aqueous micelles are similarly highly dynamic (Ulbricht and Hoffmann, 1993).

Also, micelles are equilibrium systems. What does this mean? It means that the final state, for example the average micellar size and all corresponding physical properties, are reached regardless of the pathway used to form them. The final state is thus independent of the mixing order of the components, and does not depend on the previous history of the sample. It is possible to make larger or smaller micelles by changing the environmental parameters (e.g., salt concentration) – and if two

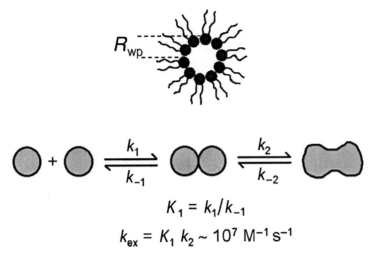

$$K_1 = k_1/k_{-1}$$

$$k_{ex} = K_1 k_2 \sim 10^7 \, M^{-1} \, s^{-1}$$

Figure 9.10 Some structural details and dynamic properties of reverse micelles: 50 mM AOT/isooctane, $w_0 = 11.1 (= 10 \, \mu l \, H_2O$ per ml), 25°C; 3.2% AOT (w/w), 1.4% H_2O (w/w); mean water pool radius: 20 Å, mean hydrohynamic radius: 32 Å; concentration of micelles: 400 μM, monomer AOT concentration; 0.6–0.9 mM; aggregation number: 125; total interfacial area: 14 $m^2 \, ml^{-1}$. (Adapted from Fletcher and Robinson, 1981, and Harada and Schelly, 1982.)

different micellar solutions are mixed with each other, the final mixed state will have an unique average dimension and shall acquire its own equilibrium state.

We have seen that aqueous micelles are important in all cases in which hydrophobic, lipophilic substances have to be solubilized – and this, as already mentioned, is the basis of the large technical importance of micelles in laundry, oil refining, cosmetics, and chemical reactivity at the oil/water interphase. Reverse micelles, as the term implies, are important in the reverse case, when a hydrophilic substance needs to be solubilized in an oily environment. Typical solvents in this case are hydrocarbons, chloroform, or CCl_4.

The addition of a moderate amount of water-soluble substances such as amino acids, sugars, or even proteins, which are insoluble in hydrocarbons, results generally in an entrapment into the water pool, and gives also rise to transparent solutions. Particularly interesting is the process by which enzymes are solubilized in a micellar solution containing as little as 1–2% water without loss of catalytic activity. The literature flourished in the 1980s on the subject of micellar enzymology, a field initiated mostly by the work at the Swiss Federal Institute of Technology in Zürich (Luisi *et al.*, 1977a, b; 1979; Luisi and Straub, 1984; Luisi, 1985; Luisi and Magid, 1986; Luisi *et al.*, 1988; and ref. therein) as well as by Martinek and Levashov's group (Martinek and Berezin, 1986, Martinek *et al.* 1978, 1981, 1986a, b) who, at that time were both in Russia. Eventually many other groups all around the world

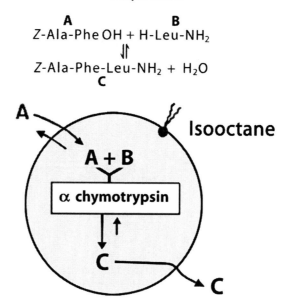

A   B
Z-Ala-Phe OH + H-Leu-NH$_2$
⇅
Z-Ala-Phe-Leu-NH$_2$ + H$_2$O
C

Figure 9.11 A case of selective compartmentation in reverse micelles, permitting the synthesis of a peptide by the reverse protease action. The product **C**, produced in the water pool, is expelled into the outside hydrocarbon environment due to its insolubility in water. (Adapted from Barbaric and Luisi, 1981.)

joined the field (Pileni, 1981; El Seoud, 1984; Hilhorst *et al.*, 1984; Han and Rhee, 1986; Waks, 1986 – see also the other many references in Luisi and Straub, 1984). In this way, enzymatic reactions, occasionally even characterized by an enhancement of the turnover number, could be carried out in practically apolar solvents – such as isooctane, decane, chloroform.

The importance of the particular compartmentation in this field is made apparent by a series of interesting and partly still unexplained effects. For example, when the amount of water is varied in the reverse micellar solution, the maximum enzyme activity – even in the case of hydrolases – is not observed with higher water-content values, but with relatively low amounts of water. In addition, the local pH – due to the constraints of the water pool – is anomalous with respect to the pH value in water (El Seoud, 1984; Luisi and Straub, 1984).

Reverse micelles are the first compartment structures for which the phenomenon of micelle self-reproduction has been described (Bachman *et al.*, 1990; 1991). This experimental work was a follow up of a theoretical study by Varela and Luisi (Luisi and Varela, 1990), and is it this that eventually brought to light the self-reproduction of aqueous micelles and vesicles. This has been covered already in Chapter 7, on the chemistry of self-reproduction.

Figure 9.11 shows a classic example of compartmentation of reverse micelles (Barbaric and Luisi, 1981): chymotrypsin entrapped in reverse micelles catalyzes

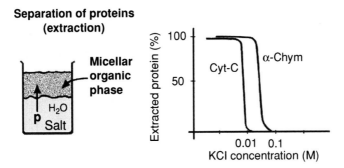

Figure 9.12 The uptake of water-soluble proteins (p) from an aqueous solution to an AOT micellar solution as a function of the salt concentration in the water phase. Note the remarkable difference between cytochrome-C (Cyt-c) and α-chymotrypsin (α-Chym), so that in principle it is possible to separate one from the other.

the peptide synthesis and since the product of the reaction is water insoluble – and soluble in the hydrocarbon – it is expelled outside of the micelles, and this drives the chemical equilibrium more completely towards synthesis.

Enzymes are large molecules with respect to the initial size of the water pool, and their inclusion in the reverse micelles brings about a re-organization of the micelles (Zampieri *et al.*, 1986). I will not dwell here on all the structural studies concerned with this issue, nor on the questions related to the thermodynamic of protein solubilization. It should be mentioned, however, that the facile solubilization of proteins in reverse micelles, and the fact that this phenomenon has some kind of specificity, has given rise to a large volume of literature on the subject (Leser and Luisi, 1989, 1990, and references therein), largely with the idea that the process could be utilized on a laboratory scale. Figure 9.12 gives one example of this specificity. As far as I know, this process has never reached the stage of a large-scale industrial or semi-industrial separation for proteins, it is nevertheless very interesting from a thermodynamic and mechanicistic point of view (Caselli *et al.*, 1988; Bianucci *et al.*, 1990; Maestro and Luisi, 1990).

Even micro-organisms can be solubilized in reverse micellar solution, thus permitting microbiology in overwhelmingly organic solvent (Häring *et al.*, 1985, 1987; Hochköppler and Luisi, 1989, 1991; Hochköppler *et al.*, 1989; Pfammatter *et al.*, 1989; 1992; Famiglietti *et al.*, 1992; 1993). Again, the percentage of water in such systems is very limited, and the reason and mechanism of this phenomenon is still largely unexplained.

Nucleic acids can also be solubilized in reverse micelles, including ribosomes and plasmids, (Imre and Luisi, 1982; Palazzo and Luisi, 1992; Pietrini and Luisi, 2002; 2004; Ousfuri *et al.*, 2005), which also gives rise to a series of interesting structural and thermodynamic questions. In particular, high-molecular-weight

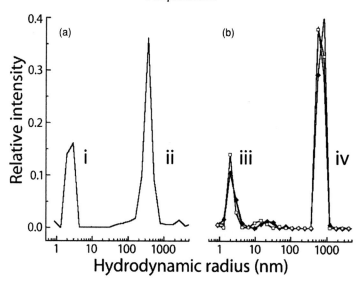

Figure 9.13 Dynamic-light-scattering size distribution (angle 120°) of a C_8PC reverse micellar solution, containing aqueous DNA solution, $w_0 = 5$. (a) 0.5 mg ml^{-1} DNA; (b) 4 mg ml^{-1} DNA. In (b) three size distributions are plotted, referring to: 15 min (—); 1 d (-◆-); 6 d (-O-) from the preparation of the micellar solution (from Ousfuri *et al.*, 2005); i and iii are empty micelles; ii and iv are DNA-containing micelles.

DNA acquires a condensed form in reverse micelles, with the characteristic "psi-spectrum" (Imre and Luisi, 1982; Pietrini and Luisi, 2004; Ousfuri *et al.*, 2005). This super-condensation of the DNA macromolecules, and their non-covalent cross-linking to yield the psi-spectrum, is due to the restricted environment.

Figure 9.13 shows one recent experiment (Ousfuri *et al.*, 2005), in which DNA has been added to a micellar solution containing an "empty" micellar water pool of only 10 nm diameter; the presence of the large DNA brings about a rearrangement of the micellar structures, producing very large aggregates (up to one μm), which are very stable, and which host the DNA. These can be very large reverse micelles, or cylindrical, tube-like long structures.

It is also interesting to recall another compartment feature of reverse micelles, found by serendipity in the search of conditions to make reverse micelles from lecithin. When traces of water are added to a hydrocarbon solution containing lecithin, a gel is formed (Scartazzini and Luisi, 1988; Luisi *et al.*, 1990), as can be seen in Figure 9.14. This organogel, as it has been dubbed, can entrap a series of different guest molecules, which gives rise to the possibility of transdermal transport (Willimann and Luisi, 1991), to an interesting chemistry in the semi-solid state (Fadnavis and Luisi, 1989; Scartazzini and Luisi, 1988) and to interesting relationships between the high viscosity and the structure of the aggregates in apolar solvents (Schurtenberger *et al.*, 1990; 1991).

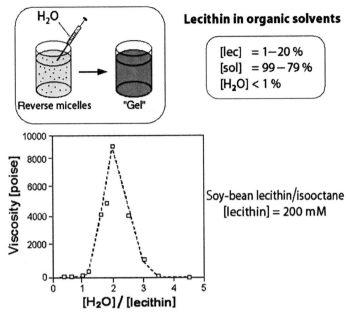

Figure 9.14 The simple addition of a minimal amount of water to a hydrocarbon solution of lecithin (or other phospholipids) brings about the formation of an organogel. (Adapted from Scartazzini and Luisi, 1988.)

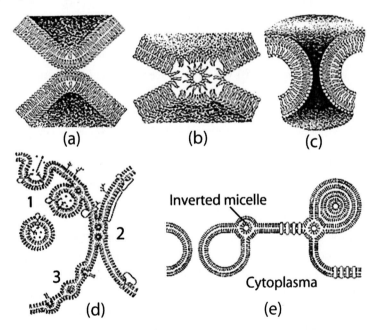

Figure 9.15 Structures like reverse micelles in vivo. Membrane fusion intermediates: (a) adhesion; (b) joining; (c) fission. (d): 1, Exocytotic fusion; 2, semi-fused interbilayer connection; 3, reverse micelles permitting enhanced permeability to Me^{2+}. (e) Aggregation of intramembranous particles in the sarcolemma after ischemia and reperfusion. (Modified from de Kruijff et al., 1980, and from Cullis et al., 1986.)

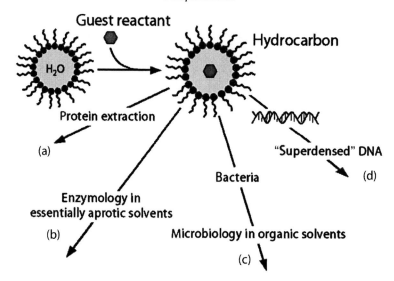

Figure 9.16 Reverse micelles as microreactors: an overview of the compartmentation behavior of reverse micelles with basic references: (a) Zampieri *et al.*, 1986; Häring *et al.*, 1988; Leser and Luisi, 1990. (b) Barbaric and Luisi, 1981; Luthi and Luisi, 1984; Walde and Luisi, 1989; Luisi and Laane, 1986. (c) Häring *et al.*, 1985; Hochköppler and Luisi, 1989; Famiglietti *et al.*, 1992. (d) Imre and Luisi, 1982; Palazzo and Luisi, 1992; Pietrini and Luisi, 2002; Ousfuri *et al.*, 2005.

It is also interesting that structures of the reverse micelle type have been observed in vivo (de Kruijff *et al.*, 1980) where they seem to absolve specific, even if not dramatically important, functions – see Figure 9.15.

An overview of the compartment properties of reverse micelles is illustrated in Figure 9.16, which also gives some of the relevant references. As already mentioned, reverse micelles are dynamic systems that rapidly exchange compartmentalized materials. There is however one limit to this: when the enclosed solutes are macromolecules. Thus, if two different populations of reverse micelles are mixed, one, say, with enzymes and the other with nucleic acids, the two macromolecules are not going to interact with each other.

There's another example of water-in-oil compartmentation, which can circumvent this problem: water-in-oil emulsions. These can be prepared by adding to the oil a small amount of aqueous surfactant solution, with the formation of more or less spherical aggregates (water bubbles) having dimensions in the range of 20–100 μm in diameter. These systems are generally not thermodynamically stable, and tend to de-mix with time. However, they can be long-lived enough to permit the observation of chemical reactions and a kinetic study.

The advantage with respect to reverse micelles lies in the fact that these "bubbles" fuse and mix with each other rather efficiently upon stirring. One example is given

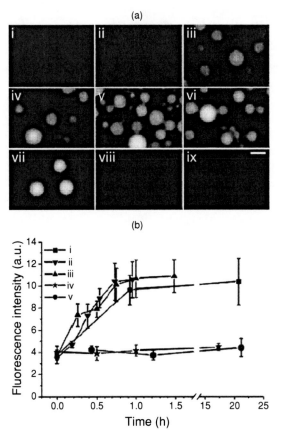

Figure 9.17 Green fluorescent protein (GFP) synthesis in water-in-oil emulsion as visualized by fluorescence microscopy. (Adapted from Pietrini and Luisi, 2004). Shown are the compartments in which GFP has been expressed (green in the original). (a) Typical micrographs of the cell-free GFP synthesis in Span 80 (0.45% v/v)/Tween 80 (0.05% v/v)/aqueous solution (0.5% v/v) in mineral oil emulsion droplets, preparation at 4 °C incubation at 37°C: (i) 0 min, (ii) 11 min, (iii) 23 min, (iv) 32 min, (v) 44 min, (vi) 57 min, (vii) 21 h. Negative control: (viii) 0 min, (ix) 21 h. The bar represents 50 μm. (b) Kinetics of the cell-free GFP synthesis in emulsion droplets, on average 10 droplets with diameters of 30–60 μm are evaluated per time point, cell-free enhanced GFP synthesis in emulsion droplets (i, ii and iii are three independent experiments) and negative control (iv and v are two independent experiments).

in the Figure 9.17, relative to one experiment in which protein expression – the green fluorescent protein – was achieved by mixing two populations of bubbles containing the complementary reagents for protein expression (Pietrini and Luisi, 2004). The subject of protein expression in surfactant compartments will be dealt with more specifically in Chapter 11.

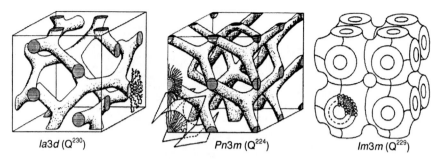

Ia3d (Q²³⁰)   *Pn3m* (Q²²⁴)   *Im3m* (Q²²⁹)

Figure 9.18 Idealized structures of hydrated didodecyl-phosphatidyl ethanolamine
showing some typical bicontinuous cubic phases. (Adapted from Seddon *et al.*,
1990; see this reference for the indicated crystallographic nomenclature.)

Cubic phases

I would also like to mention one type of non-spherical compartment that is much
less popular than micelles or vesicles, but in my view very interesting. These are the
cubic phases, so called because of their cubic symmetry. Many different types of
cubic structures have been described (Mariani *et al.*, 1988; Lindblom and Rilfors,
1989; Fontell, 1990; Seddon, 1990; Seddon *et al.*, 1990; Luzzati *et al.*, 1993).

Many single-chain amphiphiles form cubic phases when added to water in
a given composition. Two of the most well known are didodecyl-phosphatidyl
ethanolamine, and mono-olein. Figure 9.18 shows some idealized bicontinous cubic
structures of the former, including typical inverse ones. This is also highly viscous
and optically transparent as are most of the other cubic phases.

The compartmentation of cubic phases is geometrically not so well defined as
in the case of micelles or vesicles. However, several years ago the very interesting
observation was made that cubic phases can incorporate proteins up to 50% of
their weight (Ericsson *et al.*, 1983). Usually cubic phases also remain transparent
after incorporation of proteins, and in fact it has been possible to carry out circular
dichroic investigations of enzymes in such systems, (Larsson, 1989; Portmann
et al., 1991; Landau and Luisi, 1993), as shown in Figure 9.19, and even to follow
spectroscopically the course of enzymatic reactions (Portmann *et al.*, 1991).

As mentioned previously, the chemistry and biochemistry of compartmentation
in cubic phases is not as actively studied as that of liposomes and micelles; however,
these cubic phases have potential that should be explored further, mostly due to
the large capacity to incorporate biomaterials, and the peculiarity of the restricted
geometrical environment. For example the polycondensation of amino acids or other
monomers inside the lipid, chiral channels might be explored, or other reactions
where it would be advantageous to have relatively high concentrations of reagents
in a restricted tubular environment.

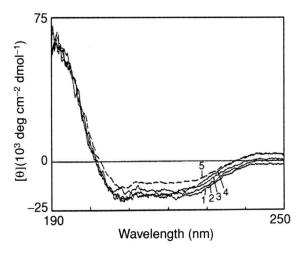

Figure 9.19 Temperature dependent circular dichroism spectra of 1.2×10^{-4} M melittin in a 43% (w/w) 1-palmitoyl-2-linoleoyl-L-3-phosphatidylcholine (PLPC), cubic phase (10 mM tris-HCl buffer, pH 7.4). Spectra taken during a heating cycle (1, 5 °C; 2, 15 °C; 3, 25 °C; 4, 35 °C; 5, 45 °C); $[\theta]$ is the mean residue ellipticity. (Adapted from Landau and Luisi, 1993.)

Size and structural properties of vesicles

Let us now look at some properties of vesicles and liposomes (liposomes can be defined as vesicles made out of lipids, although often the two terms are used synonymously). This will be a preliminary to the next chapter, where the reactivity of vesicles as models for biological cells will be considered in more detail.

Two popular vesicle-forming surfactants are shown in Figure 9.20, fatty acids and palmitoyl-oleoyl-phosphatidylcholine (POPC). In both cases, the hydrophobic parts are emphasized. Oleate, as for most long-chain fatty acids, forms vesicle spontaneously, on simple addition of its concentrated aqueous or methanol solution into water; POPC and other lipids also form liposomes spontaneously when added to water from an alcoholic solution, or by first preparing a lipid film from an organic solution (by evaporation), then adding water and stirring so as to induce a vortex.

The obtained suspension is generally a mixture of liposomes of all sizes (typically from 20 to 2000 nm), and different species are generally observed, as shown in Figure 9.21. We will see later on that "giant vesicles," spanning over a 100 μm diameter, can also be formed.

One very simple way to prepare small unilamellar vesicles is by the injection method (as shown in Figure 9.22), i.e., when vesicles are formed by adding a concentrated methanol solution of surfactant into the aqueous solution of the solute to be entrapped (Domazou and Luisi, 2002; Stano *et al.*, 2004). By this method, or by extrusion, vesicles of different average size, say 50 nm and 200 nm can

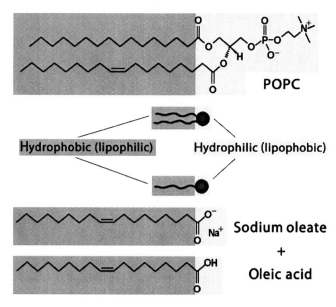

Figure 9.20 The amphiphilic character of vesicle-forming surfactant molecules; the combination of oleic acid and oleate forms vesicles in the slightly alkaline pH range.

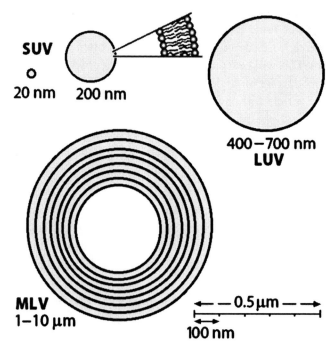

Figure 9.21 Various types of vesicles/liposomes, the so-called small unilamellar vesicles, SUV; the large unilamellar vesicles, LUV; the multilamellar vesicles, MLV.

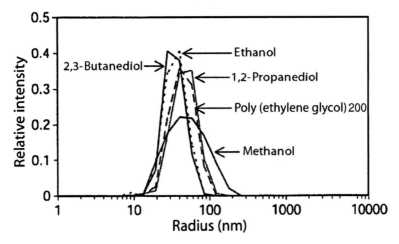

Figure 9.22 Size distribution of POPC liposomes prepared by injecting 50 μl alcoholic solution of 25 mM POPC into a 0.1 M borate buffer solution, pH 8.5: [POPC]$_{final}$ 0.5 mM, 2% (v/v) alcohol; measuring angle 90°. (From Domazou and Luisi, 2002.)

be prepared, and if the two preparations are mixed with each other, a bimodal distribution is observed. In other words, the two species, contrary to the case of micelles, do not fuse and equilibrate with each other. This is because liposomes are generally not equilibrium systems, but rather kinetically trapped systems, and the activation energy to change size is too high. This also means that in all these cases the liposomal system does not reach (does not have?) a state of absolute minimal energy.

There is another important point about the thermodynamics, and this is the fact that vesicles are generally metastable systems. Contrary to micelles, which can be indefinitely time-stable, vesicles tend to aggregate and precipitate with time. This process can, however, take many hours or days, and does not prevent reliable physical and chemical studies. This metastability is something we all in the field tend to forget too easily, also because it is actually not such a major problem. These processes are chemically irreversible, and Figure 9.23 illustrates qualitatively a situation that can occasionally occur: even in the case of a partial chemical equilibrium between monomer and aggregate, the latter aggregate M_n is slowly transformed into M'_n – e.g., into multilamellar insoluble vesicles – and the reversible transformation is not possible due to the high activation energy. In this way, the system slowly decays into the insoluble form.

Going back to the physical properties, it should be mentioned that vesicles can be seen as a microphase, which can undergo temperature-induced phase transitions. Thus, liposomes can pass from a more ordered state at low temperature (the so-called

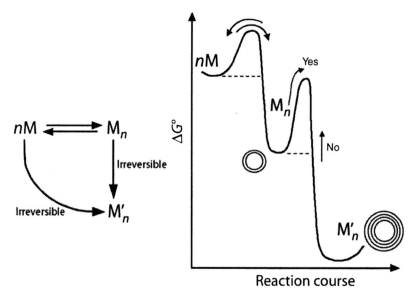

Figure 9.23 Partial equilibrium in vesicle systems, with an irreversible step leading to water-insoluble large multilamellar aggregates, see Luisi, 2001.

solid-like state) to a more disordered state at a higher temperature (liquid-like state, or liquid crystalline phase), as qualitatively illustrated in Figure 9.24. The phase-transition temperature is a function of the length of the acylated chains, and the longer the chain, the higher the transition temperature.

In particular, the transition temperature also influences the physical parameters, such as permeability and stability (liposomes can only be prepared at temperatures above T_m), and the encapsulation of solutes is rather low below this temperature (Janiak *et al.*, 1976; Machy and Leserman, 1987; Gennis, 1989).

This section may be concluded by mentioning that in addition to the "regular" aggregates such as micelles and vesicles, surfactants can assume a variety of other interesting aggregation forms. Ribbons and complex forms of helical strands, tubules, and microcylinders of various kinds have been observed. For example 5'-(1,2-dimiristoyl-*sn*-glycero(3)phospho)cytidine (DMP-cytidine) forms helical strands (Itojima *et al.*, 1992). Instead, stable vesicles can be prepared from phosphatidyl-cytidine (Bonaccio *et al.*, 1994). However, even in the case of phosphatidyl-nucleotides, small environmental changes produce the formation of helical patterns and other complex, stable, geometrical figures (Bonaccio *et al.*, 1996).

In general, it may be said that the variety of architecture built by surfactants and lipids in particular is extremely rich, that small variations in chemical structure of the surfactant may bring about significant changes in the supramolecular structure of the

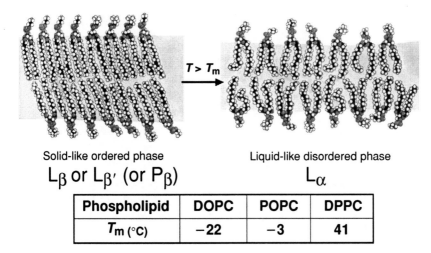

Solid-like ordered phase   Liquid-like disordered phase

L_β or $L_{\beta'}$ (or P_β) L_α

Phospholipid	DOPC	POPC	DPPC
T_m (°C)	−22	−3	41

Figure 9.24 Temperature-induced phase transitions in the bilayer of liposomes; PC = phosphatidylcholine; PO = palmitoyl-oleoyl; DP = dipalmitoyl; and DO = dioctyl; T_m represents the phase-transition temperature. (Modified from Robertson, 1983.)

aggregates, and that, due to this fine tuning, the relation between surfactant structure and supramolecular structure presents considerable difficulties. In particular it is very difficult to predict the aggregate form, particularly for a novel surfactant.

The water pool and the membrane of vesicles

The stiffness of liposomes brings about a series of problems in the chemistry of compartmentation – for example liposomes do not fuse with each other and therefore do not easily exchange material – contrary to the case for micelles. Also, vesicles are generally characterized by a very poor permeability – compounds swimming in the outside bulk water are not easily promoted inside. Figure 9.25 gives some values of the permeability of some common substances into lecithine liposomes. Even water and glycerol have great difficulty in permeating in and out the double layer. The restricted permeability of liposomes may allow a significant concentration gradient to be maintained across the bilayer, for example with small vesicles a pH difference of 5 units can be maintained (Swairjo *et al.*, 1994); and a comparably large gradient of phosphate ions (0.5 M inside and 0.05 M outside) is observed.

Conversely, once something has been entrapped, the leaking out of the solute is not a fast process. Partial leaking can however be a disturbing feature, depending on the actual conditions.

The relative rigidity and poor permeability of liposomes is not inconsistent with intense dynamics of the particles composing them. Figure 9.26 illustrates some of

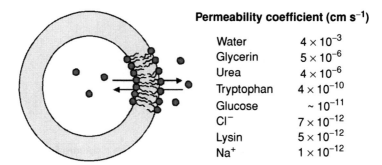

Permeability coefficient (cm s^{-1})	
Water	4×10^{-3}
Glycerin	5×10^{-6}
Urea	4×10^{-6}
Tryptophan	4×10^{-10}
Glucose	$\sim 10^{-11}$
Cl^{-}	7×10^{-12}
Lysin	5×10^{-12}
Na^{+}	1×10^{-12}

Figure 9.25 Permeability of some common substances into lecithine liposomes ($T > T_\mathrm{m}$, pH $= 7$). Generally, the permeability for polar, charged molecules and for molecules with a high molecular weight is small. Maximum permeability is when $T = T_\mathrm{m}$.

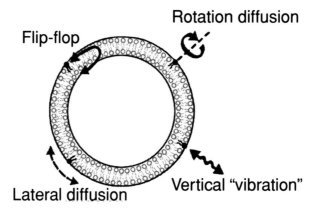

Figure 9.26 Some dynamic features of a liposome membrane (dipalmitoylphosphatidylcoline, $T > T_\mathrm{m}$). Vertical "vibration": amplitude ~ 0.3 nm; jump time 10^{-10} s. Rotation: correlation time (tc) $\sim 10^{-9}$ s. Lateral diffusion coefficient $\sim 7 \times 10^{-8}$ cm^2 s^{-1}). Flip-flop: time ~ 8 h. (Adapted from Sackmann, 1978; 1995.)

the movements of the single lipid molecules of the aggregate. Particularly important are the flip-flop movements, which in principle permit an exchange of the surfactant molecules from one layer to the other. Notice that these in-and-out movements of the lipid molecules from the bilayer should not be taken as an indication that there is a chemical equilibrium between free surfactant and aggregate: it is not that the liposomes are being continuously destroyed and re-formed (as in the case of micelles and reverse micelles), but rather that localized changes – around a given stable, kinetically trapped, supramolecular structure – are taking place.

Having considered the properties of the liposome membrane, it is now useful to look inside the water pool. We can start by calculating the internal aqueous

Table 9.2. *Geometrical properties of unilamellar POPC liposomes (from Walde, 2000, personal communication)*

Inner, outer radius (nm)	Internal aqueous volume of one liposome $(10^{-20}\mathrm{l})$	Surface of one liposome: outer, inner, total $(10^3\ \mathrm{nm}^2)$	Number of POPC molecules: outer, inner, total $(n \times 10^3)$	Liposome molarity $(\times\ 10^{-7})$ for a 10 mM POPC solution	Total internal aqueous volume ml mmol^{-1} POPC
20, 16.3	1.81	5.03, 3.34, 8.37	6.98, 4.64, 11,6	8.61	0.94
40, 36.3	20	20.1, 16.6, 36.7	27.9, 23.0, 50.9	1.96	2.37
100, 96.3	374	126, 117, 243	175, 162, 336	0.297	6.69
500, 496.3	51200	31400, 31000, 62400	4360, 4300, 8660	0.0115	35.59
5000, 4996.3	5.22×10^7	3.14×10^5, 3.14×10^5, 6.28×10^5	4.36×10^5, 4.36×10^5, 8.72×10^5	1.15×10^{-4}	360.67

volume of the liposomes, as this will be the region where reactions will take place, particularly when considering liposomes as biological cell models.

Table 9.2 shows some of these values for POPC liposomes of various radius, calculated assuming monolamellarity.

What is the physical meaning of these data? The internal radius is 3.7 nm smaller than the external one, which corresponds to the thickness of the bilayer. Liposomes with an external radius of 100 nm are made up of 336 000 lipid molecules, and in this case each compartment has an internal volume of 3.74×10^{-12} µl. This means that the total internal aqueous volume in one liter of a 10 mM POPC solution is 66.9 ml, namely 6.69% of the total volume.

If the vesicle radius is now increased to 500 nm, the internal aqueous volume becomes 512×10^{-12} µl, the total internal volume of all liposomes in a 10 mM POPC solution becomes 355.9 ml, namely 35.59% of the total volume.

Note from Table 9.2 that, in the comparison of vesicles of different sizes, the total surfactant concentration remains constant, and therefore the total surface remains constant (this is given by the number of vesicles multiplied by the total surface of one vesicle, namely the product of the third and fifth column).

These considerations are important also in view of the processes of division and/or fusion of vesicles. In particular, when a vesicle divides up, and the total surface area remains constant, the total volume must decrease. This means that water must be eliminated in the process, so as to keep the volume to surface ratio constant. Conversely, when two vesicles fuse with each other, with a constant surface area (no fresh surfactant being added), the total volume must increase to keep the volume/surface constant and water must come in. This important, characteristic feature of vesicles is represented in Figure 9.27.

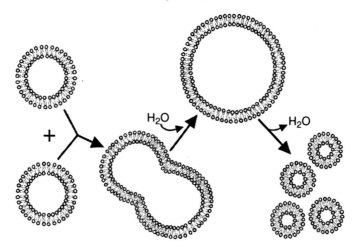

Figure 9.27 The volume to surface ratio is constant. When two vesicles fuse with each other, with a constant surface area (no fresh surfactant being added), the total volume must increase to keep the volume/surface constant and water must come in. Conversely, when a vesicle divides up, and the total surface area remains constant, the total volume must decrease. This means that water must be eliminated.

The total internal volume calculated in Table 9.2 is the volume that we have at our disposal when using compartments for biological or chemical reactions.

Prebiotic membranes

Phospholipids are the main constituents of most biological membranes, and are produced in modern cells by complex enzymatic processes. As such, they cannot be considered prebiotic compounds.

However, there have been attempts to show that they could have been formed also non-enzymatically, on the basis of prebiotic chemistry. Hargraves and Deamer were already trying to accomplish this kind of prebiotic chemistry at the end of the 1970s starting from glycerol, fatty acids and orthophosphates, or slightly more complex mixtures (Hargreaves *et al.*, 1977; Hargreaves and Deamer, 1978a, b). More recently the question of ancestral lipid biosynthesis and early membrane evolution has been re-examined (Pereto *et al.*, 2004); whereas some years back the synthesis of phosphatidylethanolamine was attempted in Oró's group (Rao *et al.*, 1987). However the question of the abiotic synthesis of glycerolphosphates cannot be considered satisfactorily solved.

A different approach is proposed by Ourisson and his group. They start from the consideration that the amphiphilic molecules in primitive membranes must have been very different from the modern eucaryotic ones (Ourisson and Nakatani, 1994), and they argue that simple polyprenyl or dipolyprenyl phosphates satisfy

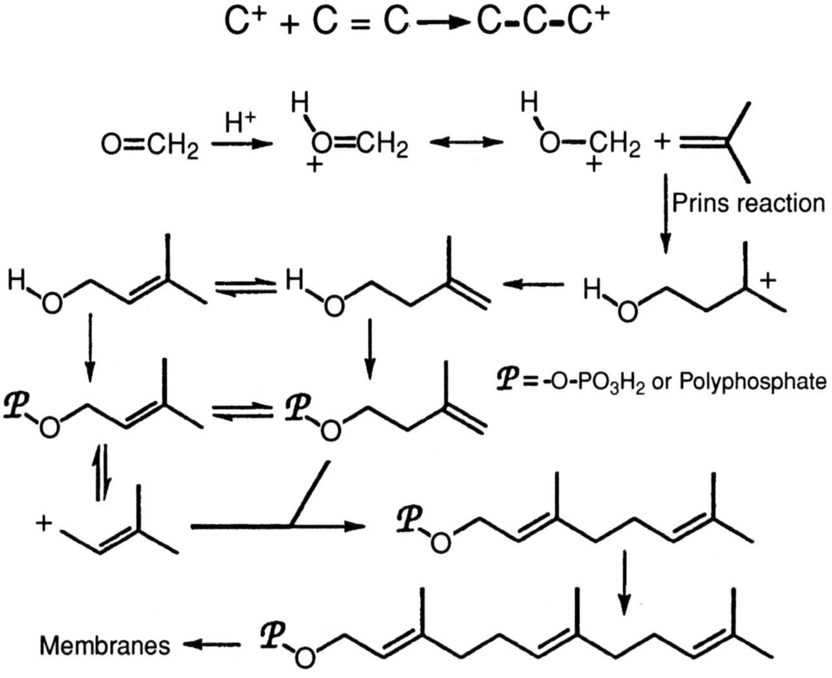

Figure 9.28 Non-enzymatic reactions leading to membranogenic polyprenyl or dipolyprenyl phosphates. (Modified from Ourisson and Nakatani, 1999.)

all conditions for being really primitive. In fact, these compounds form vesicles, provided that they contain at least 15 carbon atoms (Pozzi *et al.*, 1996). Ourisson and Nakatani (1999) also argue that it is in principle possible to synthesize these compounds starting from C_1–C_4 molecules, by reactions involving no enzyme but only acidic catalysts. The starting molecules are formaldehyde, polyphosphoric acid, which are present in a prebiotic environment (Miller and Parris, 1964; Oró, 1994), and isobutene, which appears to be present in comets, volcanic solfataric gases, and fluid inclusions in Archaean geological formation (Ourisson and Nakatani, 1999, and references therein). Figure 9.28 shows the reaction scheme proposed by Ourisson and Nakatani (1999).

More convincing evidence has been obtained for the synthesis of simpler compounds, such as straight-chain fatty acids. This observation is important because, as we have already seen in the chapter on self-organization and self-reproduction, these compounds form stable vesicles. Prebiotic synthesis of these compounds was reported for example by Nooner *et al.* (1976). More recently monocarboxylic acids have been observed from a spark discharge synthesis (Yuen *et al.*, 1981) and from a Fischer-Tropf type of reaction (McCollom *et al.*, 1999; Rushdi and Simoneit,

Sidebox 9.1

David Deamer
University of Santa Cruz, California
Prebiotic membranes
It is not yet understood how life began on Earth nearly four billion years ago, but it is certain that at some point very early in evolutionary history life became cellular. All cell membranes today are composed of complex amphiphilic molecules called phospholipids. It was discovered in 1965 that if phospholipids are isolated from cell membranes by extraction with an organic solvent, then exposed to water, they self-assemble into microscopic cell-sized vesicles called liposomes. It is now known that the membranes of the vesicles are composed of bimolecular layers of phospholipid, and the problem is that such complex molecules could not have been available at the time of life's beginning. Phospholipids are the result of a long evolutionary process, and their synthesis requires enzymatically catalyzed reactions that were not available for the first forms of cellular life.

The alternative is that the first cells simply used lipid-like molecules that were present in the environment. These can be produced by a variety of geochemical reactions, and we know that even today living organisms use this trick. For instance, vitamin A and E are lipid-like molecules that are incorporated into membranes. Could simpler amphiphiles have been present? The answer is yes. Certain stony meteorites called carbonaceous chondrites contain 1–3% of their mass as organic compounds. The organic material was synthesized in the early Solar System by non-biological processes, yet it contains a surprising array of molecules that are involved in life processes today. For instance, over 70 different amino acids have been identified. Amphiphilic molecules are also present, having hydrocarbon chains ranging up to 12 carbons long. The organic components of carbonaceous meteorites offer a guide to the sorts of organic molecules that were likely to be present on Earth four billion years ago.

Could these compounds form membranes? Again, the surprising answer is yes. When the organic compounds are isolated by organic solvent extraction and exposed to water, they assemble into membranous structures. It is known that the amphiphilic compounds present in the mix are similar to soap molecules, having both hydrocarbon chains and carboxylate groups. These also form membranous vesicles as shown in the Figure 9.29. Such vesicles are able to encapsulate other molecules, including nucleic acids and proteins.

The conclusion is that membranous vesicles readily form a variety of amphiphilic molecules that would have been available in the early Earth environment, along with hundreds of other organic species. It is likely that during the chemical evolution leading to the first catalytic and replicating molecules, the ancestors of today's proteins and nucleic acids, membranous vesicles were available in the prebiotic environment, and ready to provide a home for the first forms of cellular life.

2001). Quite recently, evidence for the fact that such compounds can be formed under prebiotic conditions has come from astrobiology: thus Dworkin *et al.* (2001) reported their synthesis in simulated interstellar/precometary ices; and amphiphilic components were observed in the Murchison carbonaceous chondrite by Deamer and his group (Deamer, 1985; Deamer and Pashley, 1989). In this regard, Sidebox 9.1 by David Deamer gives some insight into this subject, and Figure 9.29, gives pictorial evidence for the prebiotic existence of membranes.

The case of oleate vesicles

Of all mentioned prebiotic membranogenic molecules, the ones that have gained more attention in the literature are long-chain fatty acids. In addition to their prebiotic relevance, these compounds are relatively simple from the structural point of view, and most of them are easily available. We will see in the next chapter that these vesicles have acquired a particular importance in the field of the origin of life. In fact, the first investigations on self-reproducing aqueous micelles and vesicles were carried out with caprylate (Bachmann *et al.*, 1992) and most of the recent studies on vesicles involve vesicles from oleic acid/oleate (for simplicity we will refer to them as oleate vesicles). In this section, I would like to illustrate some of the basic properties of these surfactant aggregates.

Very important in this field are the pioneering studies by David Deamer and his group. Figure 9.30 shows the behavior as a function of pH, and shows that oleate makes micelles at higher pH, where the carboxylate is completely ionized; vesicles are formed at lower pH values. Quite interesting is the increase of pK_a, from a typical value of 4–5 to 8.5: a pK shift of over four units, due to the fact that the proximity of carboxylate groups makes the dissociation of a proton more difficult. This is a property that emerges from the collective behavior of the aggregate, a nice example of emergence. In fact, most of the studies on oleate vesicles have been performed in the pH range around 8.0–8.5. In principle, however, one can also work at lower pH values.

The cac (critical aggregate concentration) values for oleate are in the millimolar range, which means that at the operational concentration of 10–50 mM there will be a significant concentration of monomer in equilibrium with the aggregate. This consideration allows us to go back to the question of whether vesicles are chemical equilibrium systems. Oleate vesicles cannot be considered proper chemical equilibrium systems, however they behave in a mixed way, with some features that are typical of micelles in equilibrium (Luisi, 2001).

As apparent from Figure 9.30, there is a gradual transition from the oleate micelle region to the vesicle region, and in fact a common way to obtain vesicles is to inject a few microliters of a high-alkaline-pH sodium oleate solution into a pH

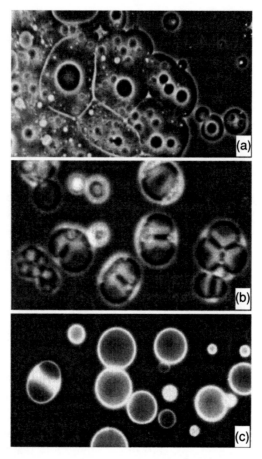

Figure 9.29 Membrane formation by meteoritic amphiphilic compounds (courtesy of David Deamer). A sample of the Murchison meteorite was extracted with the chloroform–methanol–water solvent described by Deamer and Pashley, 1989. Amphiphilic compounds were isolated chromatographically on thin-layer chromatography plates (fraction 1), and a small aliquot (~1 μg) was dried on a glass microscope slide. Alkaline carbonate buffer (15 μl, 10 mM, pH 9.0) was added to the dried sample, followed by a cover slip, and the interaction of the aqueous phase with the sample was followed by phase-contrast and fluorescence microscopy. (a) The sample–buffer interface was 1 min. The aqueous phase penetrated the viscous sample, causing spherical structures to appear at the interface and fall away into the medium. (b) After 30 min, large numbers of vesicular structures are produced as the buffer further penetrates the sample. (c) The vesicular nature of the structures in (b) is clearly demonstrated by fluorescence microscopy. Original magnification in (a) is × 160; in (b) and (c) × 400.

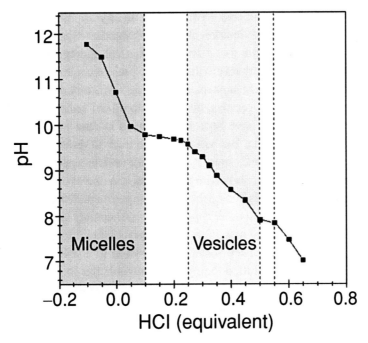

Figure 9.30 The behavior of oleate surfactants as a function of pH: equilibrium titration curve of sodium oleate at 25 °C. Note the micelles at higher pH, and the vesicles at lower pH. The chemical name of oleic acid is *cis*-9-octadecenoic acid, with 18 carbon atoms. (Modified from Cistola *et al.*, 1988.)

8–8.5 aqueous buffer. As for most long-chain fatty acids, there is a spontaneous vesiculation in the form of unilamellar vesicles, with a size distribution that depends on the preparation procedure – initial concentration, velocity of injection, pH, etc. – but typically with radius in the range 20–1000 nm.

Concluding remarks

The self-assembly of surfactants is an important phenomenon and in particular micelles and vesicles have been studied in several different fields, from chemical catalysis to drug delivery, from cosmetics to nano-technology, from industrial cleansing to household laundry. Particularly important for the aim of this book is the capability of forming compartments, and several examples with micelles, reverse micelles, cubic phases, and vesicles have been given. This point has already been outlined in the chapters on self-organization, self-reproduction, and now here. The general principle of interest has already been schematized in Figures 5.2 and 5.3. It is important to note again that the spontaneous formation of vesicles, at the cac, produces a cell-like bilayer structure, as well as determining a segregation of different kinds of solutes (ionic, hydrophobic, and hydrophilic) in the compartment. The fact that these compartments form spontaneously, and possibly with prebiotic compounds; and the fact that these spherical aggregates are capable of self-reproduction, adds remarkably to the possible prebiotic relevance of these compounds. As we will see better later on, the anionic surfactant oleate also strongly interacts with phospholipid vesicles, giving rise to mixed vesicles; and as already mentioned, oleate vesicles have been the preferred system for studying vesicle growth and self-reproduction. We will also come back to these important aspects in the next chapter, which deals more directly with vesicle reactivity.

To conclude this chapter, it is pertinent to go back to one of the arguments introduced at the beginning of this book, where the compartmentalistic approach for the origin of life is compared with other approaches, most notably the "prebiotic" RNA world. It is in fact interesting that a few important molecular biology laboratories, including groups that have been fostering the vision of the RNA world for the origin of life, have finally shifted their interest towards the chemistry of vesicles. This may prelude a convergence of research interest towards a unified strategy of seeing the two approaches (compartmentalist and RNA world) as complementary and mutually enriching, rather than opposite to each other.

Questions for the reader

1. Liposomes of different sizes can be prepared by extrusion, and when mixed with one another they remain stable in their given dimension, they do not equilibrate into one another. Why is this so? How would you prove if there is one size that corresponds to the minimum of energy?

2. In the case of mineral particles, by making them smaller and smaller, you gain a larger and larger total surface. Is this also true for the total surface of micelles and vesicles when they divide? What about the total volume?

3. Micro-organisms have been solubilized in hydrocarbon reverse micellar solutions. Would then the following experiment be possible: make reverse micelles in raw naphtha as organic solvent, have sulfur-destroying micro-organisms in the water pool of the micelles, and in this way purifye naphtha from, say, undesired sulfur containing compounds?

4. Make an organogel with lecithin and entrap an enzyme in the aqueous gel compartment. Can you make in this way a column for enzymatically induced chemical transformations?

10

Reactivity and transformation of vesicles

Introduction

Vesicles are commonly considered models for biological cells. This is due to the bilayer spherical structure which is also present in most biological cells, and to the fact that vesicles can incorporate biopolymers and host biological reactions. Self-reproduction, an autocatalytic reaction already illustrated in the chapters on self-reproduction and autopoiesis, also belongs to the field of reactivity of vesicles. Some additional aspects of this process will be considered here, together with some particular properties of the growth of vesicles – the so-called matrix effect.

Simple reactions in liposomes

Preliminary to biological studies is the incorporation of biopolymers and other reagents in the water pool of vesicles. For reasons which have been illustrated in the previous chapter , one cannot rely on the spontaneous diffusion of solute inside the liposomes, mostly due to the poor permeability of the liposome membrane. Only occasionally, depending upon the chemical nature of the solute and/or the lipid surfactant, some restricted and selective permeability is observed. In the experiment illustrated in Figure 10.1, an apolar molecule is capable of permeating inside, where it reacts with phosphate ions by opening of the ring: the product, being now polar, is trapped inside. In this case, although the permeability is very low, the irreversible chemical transformation of the reagent inside the liposomes drives the incorporation process. Another example is the slow permeation of adenonine diphosphate (ADP) inside liposomes.

Since permeation does not work for enzymes and other macromolecules, these are often incorporated by physical entrapment during the formation of vesicles, and an example is given in Figures 10.2 and 10.3, which illustrate ADP polycondensation inside vesicles. The enzyme is entrapped during the vesiculation and

214

Forced compartmentation of charged reagents

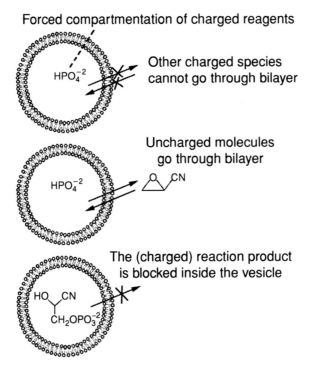

Other charged species
cannot go through bilayer

Uncharged molecules
go through bilayer

The (charged) reaction product
is blocked inside the vesicle

Figure 10.1 Reaction features of surfactant aggregates, showing an example of permeability of non-charged molecular species, and forced compartmentation of charged species. (From Luisi *et al.*, unpublished data.)

then ADP can slowly permeate from the outside to the inside of vesicles and be enzymatically polymerized into poly(A) by the entrapped enzyme (Chakrabarti *et al.*, 1994; Walde *et al.*, 1994a). This is a reaction that was investigated originally by Oparin in aggregate systems, and appropriately enough, the cited article (Walde *et al.*, 1994a) is titled "Oparin's reactions revisited."

In the case of Walde *et al.* (1994a) the synthesis of poly(A) – which can be viewed as a simple form of RNA – proceeded simultaneously with the self-reproduction of vesicles, thus providing a "core and shell" reproduction, as schematically illustrated in Figure 10.3.

There are several other ways to entrap solutes inside the liposomes, and the entrapping efficiency depends on the structure of liposomes (small unilamellar, large unilamellar, multilamellar, vesicles, etc.) and from the technique for liposome preparation (Roseman *et al.*, 1978; Cullis *et al.*, 1987; Walde and Ishikawa, 2001).

Also the surface of liposomes can be utilized for facilitating chemical reactions. One example is the liposome-aided synthesis of peptides already mentioned in Chapter 4 (Blocher *et al.*, 2000; 2001). In this case, the binding of hydrophobic

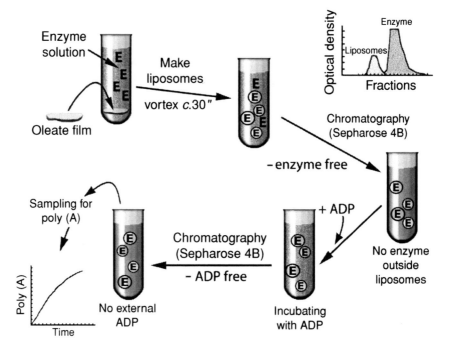

Figure 10.2 One of the procedures for entrapping an enzyme, E, (or any other solute) inside liposomes. The non-entrapped enzyme is eliminated by size exclusion chromatography, then the substrate (ADP in this case) is left to permeate throughout, the ADP excess is again eliminated chromatographically, and the reaction kinetics – due to the internalized enzyme reaction – measured. (From Walde *et al.*, 1994b.)

NCA-activated aminoacids (NCA = *N*-carboxyanhydride) to the lipophylic surface of palmitoyl-oleyl-phosphatidyl-choline (POPC) liposomes, has permitted the condensation of Trp-oligomers up to a polymerisation degree of eighteen, which is considerable, given that in water the synthesis by way of the same reaction is limited to oligomers of five to six due to their insolubility. The hydrophobic character of the liposome shell can also operate the chemical selection from a mixture of solutes, as illustrated qualitatively in the Figure 10.4: the most hydrophobic solute can in principle be selected out followed by selective polycondensation.

Going back to the entrapment into the water pool, work on enzymes in liposomes has been and is an active research field. Most of the work presented in the literature concerns the entrapment of one enzyme at a time. In particular, the excellent review by Walde and Ischikawa, 2001 provides a rich account and discussion of the various techniques used to incorporate enzymes inside liposomes, and their possible applications in chemistry, medicine, and industry. Table 10.1 is a modification and simplification of one of their tables (Walde and Ischikawa, 2001, Table 6). This

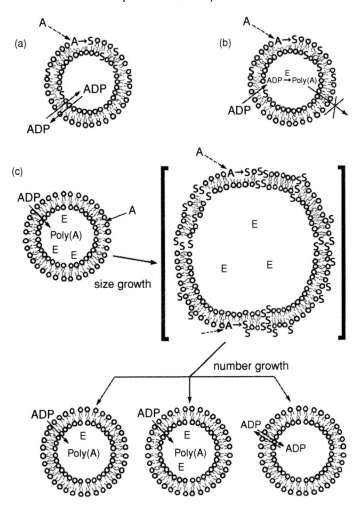

Figure 10.3 Enzymatic synthesis of poly(adenylic acid) in self-reproducing oleate liposomes (redrawn from Walde *et al.*, 1994a). (a) The ADP penetrates (sluggishly) the liposome bilayer. (b) in the presence of polynucleotide phosphorylase, ADP is converted in poly(A), which remains entrapped in the liposome. (c) Polycondensation of ADP goes on simultaneously with the self-reproduction of liposomes; (A is the membrane precursor, oleic anhydride, which, once added, induces the self-reproduction of liposomes; S, surfactant, in this case oleate, which is the hydrolysis product of A on the bilayer; **E** is polynucleotide phosphorylase).

table also describes in a nutshell the corresponding experiments and operational goals, and displays a large variety of enzymes and potential applications.

The aim of entrapping enzymes in vesicles is on the one hand to study enzymatic reactions in a restricted medium, and on the other hand to develop models for cellular

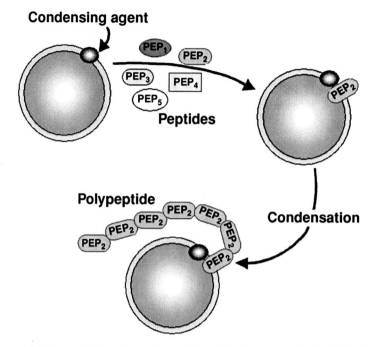

Figure 10.4 A qualitative illustration of the selective power of the hydrophobic membrane of liposomes: out of a small library of amino acids or peptides present in the external solution, only the hydrophobic ones can be picked out and polymerized by the hydrophobic condensing agent.

reactions. If aiming at models for metabolic pathways, the encapsulation of one enzyme is not enough, and actually the far-reaching goal would be the entrapment of an entire enzymatic cycle. It is fair to say that there are not significant examples of this yet in the literature, although some experiments along this line have been described. For example, there has been an attempt to entrap in one single POPC liposome the four enzymes responsible for the synthesis of lecithine starting from glycerol-3-phosphate (Schmidli *et al.*, 1991). The idea behind this project was to construct a minimal cell capable of producing its own membrane from within – an idea related to autopoiesis. This is illustrated in Figure 10.5, and more in detail in the next chapter, see Figures 11.6 and 11.7.

A quite different application of liposomes as reactive compartments is in the field of drug delivery. It was found several years ago that liposomes, because of their hydrophobic nature, strongly interact with the biological cell membrane and can actually be incorporated inside by endocytosis or other mechanisms, e.g., fusion (Allison and Gregoriadis, 1974; Gregoriadis, 1976a, b; 1988; 1995; Papahadjopoulos *et al.*, 1989).

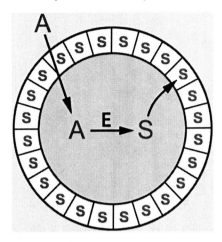

Figure 10.5 A liposome that builds its own membrane with the help of entrapped enzymes is the prototype of the simplest autopoietic minimal cell (E, A, and S as defined in Figure 10.3).

A drug entrapped in the liposomes or simply bound to it, can then be delivered inside the cell. In the cell, phosphatidylcholine (PC) liposomes can be then digested by phospholipases and thus liberate the sequestrated drugs. The therapeutical applications are somewhat restricted by the elimination of liposomes in vivo by the cellular phagocytic system. This problem can be reduced by making covalent adducts with polyethylene glycol, the so-called "pegilation," the process that gives rise to the "stealth liposomes" (Woodle and Lasic, 1992). There is a vast literature, and intense patent activity in the field, and the work by Gregoriadis is again particularly noteworthy. Several liposome preparations are already commercial or under study (Gregoriadis, 1995), including transdermal therapy (Raab, 1988; Foldvari *et al.*, 1990; Lasch *et al.*, 1991; Lasic, 1995), and cosmetics (see for example Raab, 1988; Cevc, 1992; Lasic, 1995).

It is not always necessary to incorporate the drug inside the liposomes, particularly when it is lipophylic. In this case, as already mentioned, the drug can be bound to the liposome bilayer by hydrophobic interactions. An example is given by the binding of camptothecin, an anticancer drug extracted from Chinese plants, to POPC liposomes. In particular, very small liposomes can be obtained by the injection method, which have the advantage over the very large ones of a longer circulation in vivo and a greater stability (Stano *et al.*, 2004, see Figure 10.6).

Liposomes can also be utilized for gene transport and corresponding transfection (Gao and Huang, 1995; Zhu *et al.*, 1996a, b; Reszka, 1998; Kikuchi *et al.*, 1999) – also a very active field of inquiry.

In this case the liposomes are generally dubbed with positively charged co-surfactants, such as DDAB (dimethyl-didodecylammonium bromide) or CTAB

Table 10.1. *Selected examples for enzyme-catalyzed reactions inside lipid vesicles**

Enzyme	Type of study, remarks	References
D-Amino acid oxidase	Investigation of the activity of the entrapped enzyme against externally added D-amino acids (egg PC-based MLV).	Naoi et al., 1977.
Ascorbate oxidase	Entrapped (and partially adsorbed enzyme) was active against externally added ascorbate and O_2.	Mossa et al., 1989; Annesini et al., 1992; Ramundo-Orlando et al., 1993.
Carbonic anhydrase	The entrapped enzyme mainly adsorbed onto the positively charged lipid bilayer.	Annesini et al., 1993.
	Low frequency, low amplitude magnetic fields increase the permeability of enzyme-containing lipid vesicles.	Annesini et al., 1994.
	Activity measured against externally added CO_2.	Ramundo-Orlando et al., 2000.
α-Chymotrypsin	Activity and stability measurements of the enzyme entrapped inside POPC vesicles against externally added Bz-Tyr-pNA. No activity against the larger substrate Suc-Ala-Ala-Pro-Phe-pNA or casein. Inhibition of externally present enzyme by an inhibitor protein.	Walde and Mazzetta, 1998.
	Detailed kinetic measurements and kinetic analysis (by dynamic modeling) of chymotrypsin-containing POPC vesicles.	Blocher et al., 1999.
	Fusion of chymotrypsin-containing vesicles with vesicles containing the substrate Suc-Ala-Ala-Pro-Phe-pNA by addition of partially denatured Cytochrome C.	Yoshimoto et al., 1999.
DNAase I/DNA	Entrapment of the enzyme together with the substrate (DNA) in dipalmitoylphosphatidylcholine (DPPC) lipid vesicles.	Baeza et al., 1994; 1990.
β-Galactosidase	Investigation of the change in the permeability of enzyme-containing vesicles against externally added substrate.	Annesini et al., 1994.
β-Glucosidase	Entrapment of the enzyme inside the lipid vesicles led to a stabilization of the enzyme against inhibition by externally added Cu^{2+} ions, as measured with p-nitrophenyl-D;β-D-glucopyranoside as substrate.	Sada et al., 1988; 1990.

Enzyme	Description	Reference
Glucose oxidase	A few measurements on the activity of entrapped glucose oxidase after external addition of glucose in the presence of variable amounts of added insulin.	Solomon and Miller, 1976.
	Activity measurements in egg PC-based enzyme-containing vesicles which had deoxycholate in the membrane against externally added glucose.	Ambartsumian et al., 1992.
Glucose oxidase, in combination with peroxidase or lactoperoxidase	Investigation of the activity of the two enzyme-containing anionic or cationic lipid vesicles against externally added D-glucose.	Kaszuba and Jones, 1999.
Polynucleotide phosphorylase	Experimental demonstration that the enzyme inside dimyristoylphosphatidylcholine (DMPC) vesicles was active against externally added ADP, yielding lipid vesicle trapped poly (A).	Chakrabarti et al., 1994.
	Simultaneous oleic anhydride hydrolysis resulting in a self-reproducing vesicle system.	Walde et al., 1994a.
Trypsin	Activity and stability measurements of the enzyme entrapped inside POPC vesicles against externally added Bz-Arg-pNA and Z-Phe-Val-Arg-pNA.	Graf et al., 2001.
Tyrosinase	Kinetic investigations against different externally added mono- and diphenols.	Miranda et al., 1988.
Urease	Kinetic investigation towards externally added urea. Increased stability of the entrapped enzyme.	Madeira, 1977.
Urate oxidase	Activity measurements in egg PC-based enzyme-containing vesicles which had deoxycholate in the membrane against externally added uric acid.	Ambartsumian et al., 1992.

* Adapted, with shortening and slight modification, from Table 6 of Walde and Ischikawa, 2001, with the authors' permission.

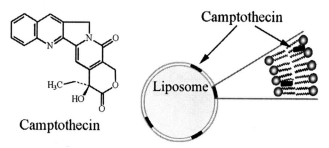

Figure 10.6 The antitumoral camptothecin (CPT), a lipophilic drug extracted from the Chinese tree *Camptotheca acuminata*, can be incorporated into the liposome bilayer due to its lipophilic character. The CPT-containing liposomes are studied as antitumor drug formulations. (Modified from Stano *et al.*, 2004.)

(cetyltrimethylammonium bromide), so as to induce an effective binding with DNA or RNA via electrostatic interactions. There is then the possibility of three different liposome/DNA species, since DNA can then be localized on the surface, or inside the water pool, or in either place. It is not yet ascertained which species is the most effective from the point of view of gene transfer, but usually the preferred operational procedure is by surface binding, as this is operationally by far the simplest one.

Giant vesicles

The so-called giant vesicles owe their name to the fact that they can reach up to 100 μm in diameter. Such over-dimensioned structures can be often observed as by-products in the normal preparation of vesicles, but they can be obtained by specific methods, for example by that dubbed electroformation (Angelova and Dimitrov, 1988).

Giant vesicles have been the subject of several international meetings and specialized literature (Luisi and Walde, 2000; Fischer *et al.*, 2000). There are several reasons for this interest. One is that, because of their size, they can be observed by normal optical microscopy, without using the much more expensive and indirect electron microscopy. Figure 10.7 shows, as an example, the transformations brought about by the addition of a water-insoluble precursor (oleic anhydride) to oleic acid giant vesicles (Wick *et al.*, 1995).

A second reason for the interest in giant vesicles is that by special micromanipulation (similar to that used in cell biology) it is possible directly and quantitatively to inject chemicals inside the compartment. An example of the effect of an enzymatic reaction inside giant vesicles is given in the Figure 10.8.

By working with giant vesicles, the chemist acquires the working habits of a cell biologist, suffering, however, from being obliged to work with only one

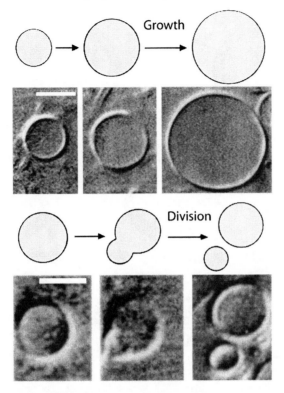

Figure 10.7 Direct observation of transformations in giant vesicles. This results from the addition of oleic anhydride to giant oleate vesicles. (Adapted from Wick *et al.*, 1995.)

compartment at a time – at variance with a preference to work with Avogadro's number of particles.

Self-reproduction of vesicles

It has already been discussed in Chapters 7 (self-reproduction) and 8 (autopoiesis) that, under certain conditions, vesicles are capable of undergoing an autocatalytic process of self-reproduction. This is a novel, dynamic aspect of the reactivity of such aggregates, which clearly has relevance for the field of the origin of life.

Vesicle self-reproduction described until now can be defined as autopoietic, since growth and eventually reproduction comes from within the structure boundary. One can also induce growth and division of fatty acid vesicles by adding fresh surfactant from the outside, for example as a micellar solution at high alkaline pH.

Let us consider the mechanism. When monomeric oleate or oleate micelles are added to a solution containing oleate vesicles, two limiting situations may occur,

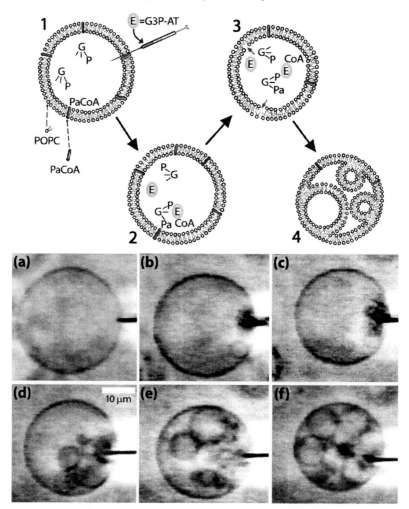

Figure 10.8 Effects of a micro-injection of *sn*-glycerol-3-phosphate acyltransferase (G3P-AT) into POPC/Palmytoyl CoA (PaCoA) giant vesicles. As shown in the scheme of the upper panel, after the injection (1), G3P-AT interacts with G3P and PaCoA, determining a partial hydrolysis of PaCoA from internal membrane wall (2). The depletion of PaCoA (3, small arrows) produces shrinkage of the vesicle, followed by the formation, on its inner surface, of smaller liposomes (4). The lower panel shows a series of phase-contrast micrographs showing the transformations of a single giant POPC/PaCoA vesicle, induced by micro-injection of 180 fl of G3P-AT solution (250 μg ml^{-1}). (a)–(f): Phase-contrast micrographs, taken 0, 10, 20, 50, 140 and 300 s, respectively, after the injection of G3P-AT into the vesicle. (Adapted from Wick and Luisi, 1996, with some modifications.)

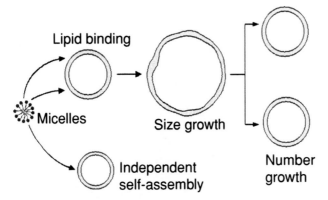

Figure 10.9 Addition of fresh surfactant (or a micellar solution thereof) to a solution containing pre-existing vesicles can follow two alternative (not exclusive) pathways: either formation of new vesicles; or binding to the existing vesicles, which may bring about growth and division phenomena (a highly idealized case is shown, with vesicles dividing in two).

as illustrated in Figure 10.9: either the added substrate molecules self-assemble by their own course, ignoring the pre-existing vesicles; or, instead, they bind to the pre-existing vesicles. These then may grow and eventually divide.

It is clear, as shown in Figure 10.9, that the choice between the two pathways is the result of a competition between the velocities of the two processes, regulated by the relative rate of binding, and assembly formation, respectively. If the rate of vesicle formation is per se very high, there is no possibility for binding and for the eventual growth and reproduction of the "old" vesicles.

For example, if fresh POPC from a methanol solution is added to a POPC liposome solution, there will be an immediate formation of fresh liposomes. In fact, the cmc in this case is of the order of 10^{-10} in favor of the aggregates over the free monomers, and the rate of formation is correspondingly extremely high. Thus, we are in the presence of an independent self-assembly mechanism.

In contrast to this is the addition of oleate surfactant – in the form of micelles or free monomer – to oleate or to POPC vesicles. In this case, the ratio of the two competitive rates is such that a considerable binding of the added fresh surfactant to the pre-existing vesicles takes place. The efficient uptake of oleate molecules by POPC liposomes (Lonchin *et al.*, 1999) as well as to oleate vesicles (Blöchiger *et al.*, 1998) is well documented in the literature.

How can some light be shed onto the mechanism?

One way is to label the pre-existing vesicles, and then follow the destiny of the label in the vesicle size distribution. The label that has been used to this aim is ferritin, which has been entrapped into vesicles. Ferritin is an iron-storage protein in plants and mammals, and consists of a hollow protein shell of *c.* 12 nm containing

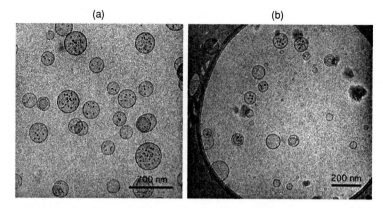

Figure 10.10 Transmission electron micrograph of ferritin entrapped in POPC liposomes (palmitoyloleoylphosphatidylcholine). Cryo-TEM micrographs of (a) ferritin-containing POPC liposomes prepared using the reverse-phase evaporation method, followed by a sizing down by extrusion through polycarbonate membranes with 100 nm pore diameters ([POPC] = 6.1 mM); and (b) the vesicle suspension obtained after addition of oleate to pre-formed POPC liposomes ([POPC] = 3 mM, [oleic acid + oleate] = 3 mM). (Adapted from Berclaz *et al.*, 2001a, b.)

in its center an iron core of *c.* 7.8 nm. This very dense iron core gives rise to a strong scattering contrast, which facilitates detection in electron microscopy. Once ferritin is entrapped in vesicles, vesicle suspensions can be frozen (vitrified) as thin aqueous layers and examined at low temperature by transmission electron microscopy (TEM). This technique also permits the determination of vesicle size and lamellarity (Böttcher *et al.*, 1995), and the number of ferritin molecules in each vesicle can even counted – see Figure 10.10.

How this ferritin label can be used for elucidating the growth mechanism is illustrated in Figure 10.11, and is conceptually simple. The distribution of sizes and ferritin content before and after addition of the fresh surfactant can be measured by TEM. Compare now Figures 10.9 and 10.11: if the fresh surfactant does its own thing and does not interact with the pre-existing ferritin-containing vesicles, the same distribution of ferritin-containing vesicles will be found at the end as at the beginning. If instead the added surfactant interacts with the pre-existing ferritin-containing vesicles, the distribution of ferritin-containing vesicles will be changed, in a way that reflects the mechanism of growth.

These considerations have given rise to an intensive TEM investigation (Berclaz *et al.*, 2001a, b). As an example, Figure 10.12 shows the size distribution of "empty" (no ferritin) and filled vesicles at the start; it also shows the size distribution of empty and filled vesicles after oleate addition; and finally the comparison between filled vesicles before and after addition. It is clear that there is a significant growth of

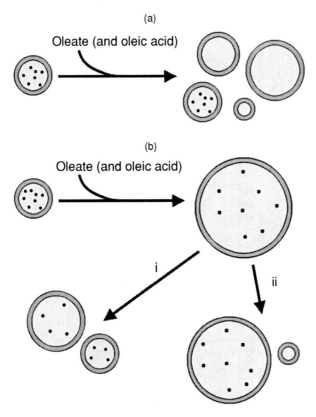

Figure 10.11 The use of ferritin as a label for the mechanism of growth of vesicles (adapted from Berclaz *et al.*, 2001a; b). Schematic representation of the possible vesicle formation and transformation processes when oleate, and oleic acid, are added to pre-formed vesicles which have been labelled. (a) The situation if only *de novo* vesicle formation occurs. (b) Growth in size of the pre-formed and labeled vesicles which may lead to division, either yielding vesicles that all contain marker molecules (case i, a statistical redistribution of the ferritin molecules) or also yielding vesicles that do not contain markers (case ii). Compare all this with Figure 10.9.

the vesicles under the given conditions: in particular a population of vesicles with dimensions larger than 200 nm, and extending up to 500 nm, has been formed. This is direct evidence of the growth mechanism depicted in Figure 10.11(b).

However, this is not all: as shown in the Figure 10.13, under certain conditions there is a small but significant concentration of small ferritin-containing vesicles of sizes that were not present before the addition of oleate. This clearly shows the process of division, as illustrated in Figure 10.11(b), process i.

Can direct evidence of the splicing process be obtained by electron microscopy? The answer is positive, as shown in Figure 10.14, the micrographs obtained utilizing

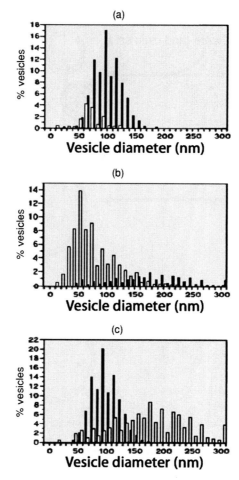

Figure 10.12 Number-weighted size distributions as obtained by cryo-TEM (adapted from Berclaz *et al.*, 2001a, b). (a) Distribution for the pre-formed POPC vesicles ([POPC] = 1.9 mM). (b) Distribution for the vesicle suspension obtained upon addition of oleate to pre-formed ferritin-containing POPC vesicles ([POPC] = 0.2 mM; [oleic acid + oleate] = 5 mM). Empty (▫) and ferritin-containing (■) vesicles are represented individually in the histogram. (c) Direct comparison of the number-weighted size distribution of the pre-formed POPC vesicles, which contained at least one ferritin molecule (■) with the number-weighted size distribution of the ferritin-containing vesicles obtained after oleate addition to pre-formed POPC vesicles (▫). Note that the total of all ferritin-containing vesicles was set to 100%.

samples that were collected and frozen at the very beginning of the process of oleate addition (Stano *et al.*, unpublished data).

In conclusion then, there is direct evidence that vesicles grow in size when fresh surfactant is added; and direct evidence of division processes. All this is of course

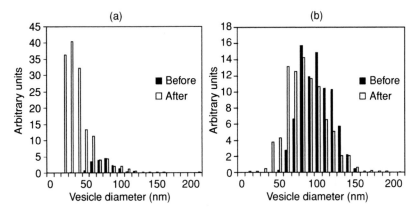

Figure 10.13 Demonstration of the process of vesicle division upon addition of fresh oleate surfactant to ferritin-labeled pre-existing POPC liposomes. (Adapted from Berclaz *et al.*, 2001a, b). Comparison of the "absolute" number-weighted size distribution (a) of the empty and (b) filled pre-formed POPC liposomes ([POPC] = 6.1 mM; ■) with the vesicles obtained after addition of oleate ([POPC] = 3 mM, [oleic acid + oleate] = 3 mM; □).

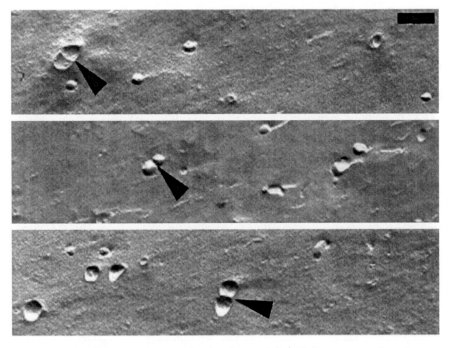

Figure 10.14 Electron micrographs showing vesicle splicing process after oleate addition. (From Stano *et al.*, in press.)

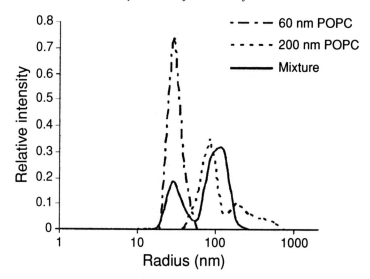

Figure 10.15 Intensity-weighted size distribution of POPC in 0.2 M bicine buffer solution, pH 8.5; 0.5 ml solution of 1 mM 30 nm extruded POPC vesicles mixed with 0.5 ml solution of 1 mM 200 nm extruded POPC vesicles. The measuring angle is 90°. (From Cheng and Luisi, 2003.)

very important if vesicles are regarded as models for biological cells – the argument that will come next.

The growth of fatty acid vesicles has been re-investigated by Chen and Szostack (2004) who, by the use of stopped-flow and fluorescence resonance energy transfer (FRET) techniques, provided interesting insights into the kinetics of this process.

The process opposite to vesicle division is that of fusion, when two or more vesicles come together and merge with each other, yielding a larger vesicle. As outlined in the previous chapter, vesicle fusion is generally not a spontaneous process. If two populations of POPC liposomes with different average dimensions are mixed with each other, they do not fuse to produce a most stable intermediate structure – they stay in the same solution as stable, distinct species. This is connected to the notion of kinetic traps, as discussed previously, and is supported by theoretical and experimental data from the literature (for example, Hubbard *et al.*, 1998; Olsson and Wennerstrom, 2002; Silin *et al.*, 2002).

This point has recently been re-investigated on the basis of dynamic light scattering, and Figure 10.15 shows the coexistence of two POPC liposome species obtained by extrusion; in order to obtain liposomes with a radius of *c.* 90 nm, extrusions through larger pores (200 nm radius) were used. The two species coexist for days in this way without interacting with each other; and the same finding is obtained with other dimensions (Cheng and Luisi, 2003).

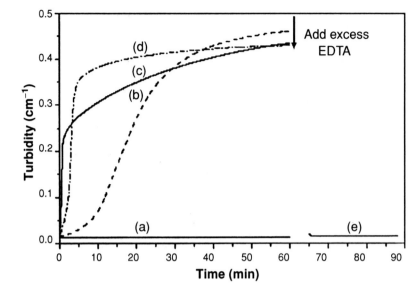

Figure 10.16 Effect of Ca^{2+} on the turbidity change upon mixing oleate vesicle solutions: 0.25 ml 1 mM 60 nm radius extruded oleic acid vesicles + 0.25 ml 1 mM 200 nm radius extruded oleic acid vesicles + 1.5 ml bicine buffer. Calcium ion concentration: (a) 0 mM; (b) 1 mM; (c) 2.5 mM; (d) 5 mM; (e) added excess EDTA to (d). (Adapted from Cheng and Luisi, 2003.)

There have been many reports in the literature of attempts to induce fusion, but actually these are rather confused, as the term "fusion" is used in a rather undiscriminate sense. I believe that the term fusion should be restricted to those processes where there is exchange and mixing of the water pools and formation of a new vesicle species. Instead, the word fusion is often used to indicate the formation of a complex between vesicles, without making clear what happens afterwards. Afterwards, there can be partial exchange of solutes or not; and often the two vesicles may depart from each other and things return to more or less the initial situation. This cannot be defined as fusion. In the literature, the use of Ca^{2+} is often suggested to induce fusion – but also in this case it is not always clear what really happens. One illustration of the addition of calcium as a non-fusion process is given in Figure 10.16, showing the effect of Ca^{2+} on negatively charged oleate vesicles. One might expect that in this case the metal cations favor fusion. At first sight, this is the case: as shown in Figure 10.16, there is an increase of turbidity when calcium is added to the vesicles, suggesting an increase of the molecular mass. However, when EDTA is added to this solution, the turbidity value is reduced to the initial value, showing the reversibility of the aggregation phenomenon: no fusion. To confirm that real fusion does not take place in this experiment (and similar ones), fluorescence techniques have been employed, with the typical terbium/dipicolinate

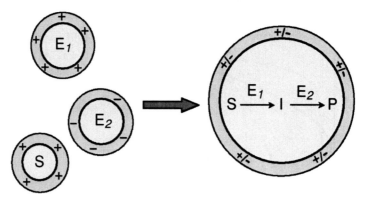

Figure 10.17 Fusion of vesicles as a way to foster reactivity and to increase the molecular complexity of the water-pool content: S, I and P are enzymatic substrate and reaction products. This is also a method to circumvent the problem of substrate permeability in liposomes. It can be seen as a model of synthetic symbiogenesis.

(DPA) assay (Wilschut *et al.*, 1980). The Tb^{3+} ion is encapsulated in one vesicle species, and DPA in the other. The mixing would result in a large increase of fluorescence (observed in the control) if the two chemicals came into contact. Nothing of this sort is observed, confirming that the two vesicle species aggregate without exchanging their water-pool material.

One efficient way to bring two vesicles species to fusion is by the use of opposite charges in the membrane, as already observed by a number of authors (Uster and Deamer, 1981; Kaler *et al.*, 1989; Kondo *et al.*, 1995; Yaroslavov *et al.*, 1997; Marques *et al.*, 1998; Pantazatos and McDonald, 1999). This is schematised in Figure 10.17. However, it is not at all clear whether such an ideal situation is really encountered in the experiments described until now.

This kind of process is interesting in several respects. It is a way to induce reactivity between the solutes entrapped in two different vesicle species. Fusion between vesicles is also a way to increase the molecular complexity of the incorporated species: for example, one can bring together enzymes and nucleic acids, or more enzyme species in order to induce, in principle, a metabolic cycle, etc.

Consider also that this kind of increase of molecular complexity due to fusion corresponds loosely to a kind of symbiogenesis, which is reminiscent of the prebiotic scenario suggested by Dyson for bringing together the world of nucleic acids and the world of proteins (Dyson, 1985). This is in turn reminiscent of the classic ideas of Lynn Margulis for symbiogenesis at the cellular level (Margulis, 1993).

All this represents work in progress, and to date the thermodynamic and kinetic aspects of this fusion process have not yet been studied.

One such surprise has been observed recently in a series of experiments in which negatively charged oleate vesicles were mixed with the positively charged DDAB

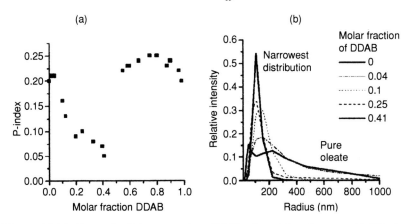

Figure 10.18 Dynamic light scattering (DLS) of vesicle mixtures. (a) P-index phase diagram and (b) size distributions (from DLS) for DDAB–oleate mixtures, total concentration 1 mM in 0.2 M borate buffer at pH 8.5, 25.0 °C, scattering angle 90°. (From Thomas and Luisi, 2004.)

vesicles. Both these surfactants display spontaneous vesiculation, each forming a broad distribution of sizes ranging from *c*. 20 to 1000 nm.

Now, when these two species are added to each other, in a given relative concentration, a new species appears with a much *narrower* size distribution. This is shown in Figure 10.18, where the P-index (a measure of the polydispersity) is plotted against the molar fraction of DDAB. The P-index drops from the initial value of 0.20 (a very broad distribution) to 0.04, a very narrow distribution (stable for months), at a relative percent of 0.4 DDAB to 0.6 oleate (Thomas and Luisi, 2004). Between DDAB molar fractions of 0.41 and 0.60, flocculation occurs, which indicates a thermodynamic instability, in agreement with other cationic systems (Kaler *et al.*, 1989; Marques *et al.*, 1998; Kondo *et al.*, 1995).

The very peculiar molar ratio 0.4 DDAB to 0.6 oleate, which gives rise to the narrow size distribution, is really noteworthy. This molar ratio corresponds closely to electroneutrality (this is not at 50:50 molarity, due to the relatively high p*K* of oleate carboxylate in the bilayer) and suggests that small mixed vesicles with an approximately equal number of positive and negative charges may enjoy particular stability. More detailed studies are needed, and this indicates the richness of the unexplored in the field of vesicles. This is shown in its fullness in the next section on the "matrix effect," which is also an unexpected phenomenon and one that may have implications for the origin of early cell.

The matrix effect

Vesicle growth and reproduction caused by addition of fresh surfactant to a solution of pre-existing vesicles has been described previously. Closer investigation of the

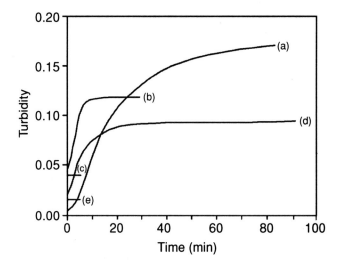

Figure 10.19 Effect of pre-added vesicles on the formation of oleic acid/oleate vesicles. Turbidity measured at 500 nm (1 cm path length) is plotted as a function of time, $T = 2\ °C$. (a) 62 µl of 80 mM aqueous sodium oleate was added to 2.438 ml of 0.2 M bicine buffer, pH 8.8 ([oleic acid/oleate] = 2 mM). (b) 62 µl of 80 mM aqueous sodium oleate was added to 2.438 ml of a 2 mM oleic acid/oleate "100 nm vesicle" suspension (0.2 M bicine buffer, pH 8.8 ([oleic acid/oleate] = 4 mM). (c) Turbidity of 2 mM oleic acid/oleate "100 nm vesicles". (d) the same as (b), but using a "50 nm vesicle" suspension. (e) Turbidity of 2 mM oleic acid/oleate "50 nm vesicles". (Modified from Blöchiger *et al.*, 1998.)

kinetics of such a process, as well as the study of the corresponding size distribution, gives surprising findings.

Consider the experiment illustrated in Figure 10.19, which shows oleate vesicle formation when an aliquot of concentrated surfactant is added to water; compared to the situation in which the same amount of surfactant is added to a solution containing pre-formed vesicles. In the second case, the formation of vesicles is remarkably accelerated, as if in the presence of a strong catalytic effect: whereas over one hour is needed to reach the turbidity plateau for oleate addition to water, the plateau is reached in less than ten minutes, curve (b), in the second case.

Note also that the kinetic progress depends on the size of the pre-added vesicles. This "catalytic" effect is present also in a ratio 1:100, or less, between the pre-added and added surfactant, reinforcing the analogy with a catalytic effect.

It appears, therefore, that the presence of vesicles accelerates the formation of "new" vesicles. It is not easy to rationalize how and why. What comes to mind is a general observation from the field of surfactants, that pre-organization makes the organization of further material easier. For example, there is no spontaneous vesiculation when POPC is simply added to water, and no significant amount of vesicles is formed by stirring so as to produce a vortex, ("vortexing"). However, when there is a pre-existing film of POPC in the flask (namely, a pre-organized surface), simple vortexing of the water suspension is enough to produce a large amount of POPC liposomes.

However, there is another important observation arising from Figure 10.19: this is the height of the turbidity plateau. Note in fact that in the experiment described by curve (b), there is the double quantity of surfactant compared to the experiment represented by curve (a); one would have expected a turbidity plateau of roughly double in (b) with respect to (a). The contrary is true: the turbidity plateau is lower in the solution with more surfactant. How can this be explained?

We may start from the recognition that turbidity is sensitive not only to concentration, but also to the size of the particles. Thus, the most likely interpretation of Figure 10.19 is that curve (b) corresponds to a much narrower size distribution. This is in fact what studies of dynamic light scattering show. Two corresponding, typical experiments are shown in Figure 10.20a, for oleate addition to pre-formed POPC liposomes, and Figure 10.20b, showing oleate addition to pre-formed oleate vesicles. The surprising result in both cases is that the size distribution of the newly formed vesicles is extremely close to that of the pre-formed ones. It is as if there were a template effect, a kind of stamp, that makes the new vesicles more or less equal to those that are already present in the solution. (The term "template effect" – rather than matrix effect – might also have been appropriate, except that this term is generally used in connection with macromolecular primary sequences with an information content).

Note that the dynamic light scattering data in Figure 10.20 report the light-scattering intensity, which is a weight average. In terms of number average, the number of particles after addition is larger than the initial one (since the concentration has doubled and the size of the particles is more or less the same). These numbers have been considered in detail in the original paper (Rasi *et al.*, 2003), and in particular it was calculated that in the case of the oleate addition to POPC (where the radius actually shifts slightly towards smaller values) the number of particles becomes more than double, whereas for oleate addition to oleate vesicles, the increase of the particle number is somewhat less than two, as there is a shift of the average radius towards larger values. More details on this phenomenology, in terms of the effects of relative concentration of the reagents, of the size of the pre-added vesicles, and of the methods of addition, are given in the literature (Lonchin *et al.*, 1999; Berclaz *et al.*, 2001a and b; Rasi *et al.*, 2003; 2004; Luisi *et al.*, 2004; Stano *et al.*, unpublished data).

It has also been shown that the matrix effect is not limited to a first addition of fresh surfactant, but can be repeated several times, thus increasing the number of particles of the same size distribution up to an order of magnitude (Rasi *et al.*, 2003;

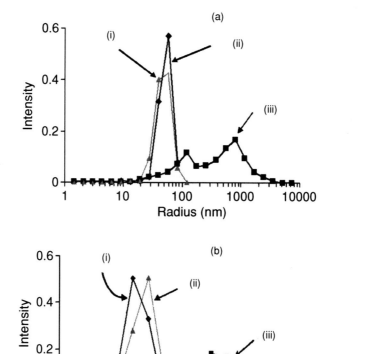

Figure 10.20 (a) Matrix effect for oleate addition to pre-formed POPC liposomes. In this case, mixed oleate/POPC vesicles are finally formed. Note the extraordinary similarity between the size distribution of the pre-formed liposomes and the final mixed ones. By contrast, the size distribution of the control (no pre-existing liposomes) is very broad. (i) Sodium oleate added to POPC liposomes, radius = 44.13, P-index = 0.06; (ii) POPC liposomes, radius = 49.63, P-index = 0.05; (iii) sodium oleate in buffer, radius = 199.43, P-index = 0.26. (b) matrix effect for the addition of fresh oleate to pre-existing extruded oleate vesicles. In this case, the average radius of the final vesicles is *c.* 10% greater than the pre-added ones, and again the difference with respect to the control experiment (no pre-added extruded vesicles) is striking. (i) Oleate vesicles extruded 100 nm, radius = 59.77, P-index = 0.06; (ii) oleate added to oleate vesicles, extended 100 nm, radius = 64.82, P-index 0.09; (iii) sodium oleate in buffer, radius = 285.88, P-index = 0.260. (Modified from Rasi *et al.*, 2003.)

Stano *et al.*, in press). The addition of fresh surfactant can also be done with a continuous reactor.

The effect is not limited to oleate, as DDAB also when added to itself or to POPC liposomes is characterized by a strong matrix effect (Thomas and Luisi, 2004). The matrix effect was re-investigated by Szostak's group, who confirmed the basic findings (Hanczyc *et al.*, 2003).

The matrix effect is thus a way to reproduce a vesicle population of a given size distribution. In an origin of life scenario, the constancy of size during self-reproduction is probably important, as it would have ensured a constancy of physico-chemical and biological properties over various generations.

In the case of the matrix effect, contrary to the autopoietic experiments described earlier, there is no need of water-insoluble precursors – it is the very addition of the same surfactant to an already existing family of vesicles that brings about the multiplication of the same size distribution. All that is needed is an initial narrow distribution of vesicles, and a continuous addition of fresh surfactant. Methods to obtain narrow size distributions in the case of spontaneous vesiculation have been described (Domazou and Luisi, 2002; Stano *et al.*, in press). In fact a prebiotic scenario may be conceived where the fresh surfactant is continuously synthesized *in situ*, and thanks to the matrix effect the same sizes are propagated over and over again. Of course there is no way to demonstrate that this is what really happened in prebiotic times – it is fair, however, to claim that, given the simplicity of the process, there is a reasonable probability that a process of this sort may have occurred (Luisi *et al.*, 2004).

The importance of size for the competition of vesicles

The uptake of oleate by pre-added vesicles, and in particular the matrix effect, permits regulation of the growth of the size and the number of particles, and in this way it is possible to tackle a series of novel questions. One such question is: "is there a difference in the rate of uptake of fresh surfactant between two vesicle populations of different sizes?"

This question is interesting because the relative rate of growth and self-reproduction may correspond to a competition between two kinds of populations. In particular, this may simulate the competition between two different kinds of organisms coexisting in the same medium, whereby the added surfactant can be seen as the nutrient the two organisms are in competition for.

Experiments have been set up in this direction (Cheng and Luisi, 2003) utilizing extruded oleate vesicles having radii of 31.6 and 64.1 nm. For the same surfactant concentration, the total surface areas are the same for the two families of vesicles (neglecting the differences arising in aggregation number due to the differences

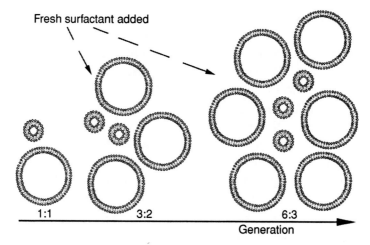

Figure 10.21 Larger vesicles grow faster than smaller ones: a starting ratio 1:1 between larger and smaller vesicles becomes 3:2 and then 6:3 in the following generations (idealized). (Modified from Cheng and Luisi, 2003.)

between the inner and outer leaflets). This also means that for the same total concentration, the number of vesicles is much larger in the case of the family with the smaller radius. Generally, in this kind of competition experiment, care should be taken to work either with a constant number of particles, or with a constant surface area. Some typical experiments, carried out under conditions where the vesicle number concentration is roughly the same for the 31.6 and the 64.1 nm families, have shown that the larger vesicles grow faster (by roughly a factor of three).

The greater reactivity of the larger vesicles can be expressed by Figure 10.21, which illustrates qualitatively the outcome of competition during generations of self-reproduction. Since larger and smaller vesicles are present in the same number, the larger reactivity of the larger vesicles could be explained by the fact that their total surface and therefore the uptake area is larger. This is the easy explanation. On the other hand, it might have been expected that the smaller vesicles, characterized by a greater curvature radius, might have relaxed more eagerly into a more relaxed state by absorbing more rapidly the added material.

The issue of the competition between vesicles of different sizes has been also examined by Szostak's group in the very interesting context of the incorporation of RNA (Chen *et al.*, 2004). These authors argue that RNA encapsulated in fatty acid vesicles exerts an osmotic pressure on the vesicle membrane that drives the uptake of additional membrane components, leading to membrane growth at the expenses of relaxed vesicles, which shrink. Although the effects are rather small, the authors argue that this difference in growth may have implications on the Darwinian

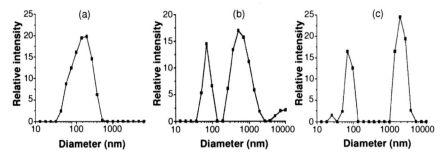

Figure 10.22 Vesicle size selection by RNA added to a mixture of small (~80 nm diameter) and large vesicles (~160 nm) from 0.5 mM POPC–3.5% CTAB in 20 mM sodium phosphate buffer (pH 7.0). (a) The initial size distribution of the 1:1 mixture; (b) the size distribution 4 min after RNA addition, during the selection process; (c) the stable final size distribution after 15 min, with one peak for small diameters (~90 nm) and a second peak indicating aggregates of large vesicles (>1000 nm). As mentioned in the text, this process is reversible upon addition of RNAase (Thomas and Luisi, 2004.)

evolution of early cells (Chen *et al.*, 2004). Interesting here is the argument that there is a relation between physical factors on the membrane and evolution.

In a different kind of experiment, CTAB vesicles of different sizes were used to study the binding interaction with t-RNA (Thomas and Luisi, 2005, see Figure 10.22). In this case, the work was prompted by the idea of investigating whether RNA might present some kind of specific interaction with phospholipid bilayers. To this aim, vesicles from POPC containing 3.5% molar fraction of the positively charged single-chained surfactant CTAB and the t-RNA mixture were used. Two different populations of narrowly sized extruded vesicles having an average radius of 40 and 80 nm, respectively, were used to study the possible influence of size on the interaction.

The RNA was then added to the vesicle solution in separate experiments. The size distribution of the vesicles before and after the addition of t-RNA was investigated by dynamic light scattering and by direct ultraviolet (UV) optical density observation.

Surprisingly, it was observed that the aggregation behavior of t-RNA to the charged vesicles was strongly dependent on the vesicle size – although in this experiment the size of the two vesicles differs only by a factor of two. In particular, the larger vesicles aggregated rapidly upon RNA addition, whereas the smaller ones did not, although they too were equally capable of binding RNA. The aggregation of the larger vesicles was completely reversible: as soon as RNAase was added to the aggregates, the initial size distribution was obtained again – which showed that the aggregation process occurred without significantly affecting the vesicular properties.

In conclusion, larger vesicles from POPC/CTAB readily aggregated in the presence of RNA in a completely reversible way; smaller vesicles – with the same chemical composition and only half as small – did not. The question arising from these data is whether the inherent ability of RNA to discriminate between different vesicle sizes with such a fine tuning might have been important with regard to early cell evolution. This connects with the illustration in Figure 10.21, showing that different sizes can grow and reproduce with different efficiency, as well with the previously mentioned observation of Chen *et al.* (2004).

There is a quite different field of research where there is an interaction between the world of nucleic acids and the world of vesicles. This is at the level of the phosphatidyl nucleotides. This is a new family of amphiphiles, where the phospholipid group is connected to a nucleobase. These compounds have been shown to be capable of forming liposomes (Bonaccio *et al.*, 1994a, b; Bonaccio *et al.*, 1996). These liposomes display the recognition chemistry of nucleic acids, but in the spherical compartmentalized structure of vesicles. The idea behind this research can easily be understood: prepare "complementary liposomes," namely liposomes with adenine groups attached to the phosphatidyl moiety; and a distinct family of liposomes with guanine (or guanine with thymine), mix them with each other, and see whether the recognition also works at the level of liposomes. When this project was started, we in our research group were rather excited at the possibility of having the two complementary families fuse with each other and form liposomes with two complementary leaflets, held together by Crick–Watson base pairs . . . Nucleic-acid recognition at the level of vesicles?

Nothing of this sort really happened. Or rather, some kind of complementary recognition was detected, but in a frustratingly weak manner; and studies with monolayers were not particularly more successful (Berti *et al.*, 1998, 2000). The reason for this failure, as evidenced by an NMR investigation carried out by Anna Laura Segre's group in Rome (Bonaccio *et al.*, 1997), lay mostly in the fact that the nucleobases attached to the phospholipids did not like to "swing" in water – they were mostly tacked in the lipophylic bilayer and were not available for "talking to" the complementary nucleobases. Some interesting results did emerge from these studies, for example concerning the morphology of these ampliphilic aggregates (Bonaccio *et al.*, 1996), but not what we had hoped. Research does not go always the way you want . . .

Concluding remarks

The fact that, working with these liposome systems, new things are always being discovered, is pleasant, but it also reveals how little we know about their thermodynamic and kinetic properties. It would have been impossible to predict the matrix effect or the influence of vesicle size on certain rate processes. Perhaps these emergent properties cannot be foreseen because the systems are too complex.

One inherent difficulty, with vesicles, is that they, unlike micelles, are not chemical equilibrium systems – each vesicle species is generally its own kinetic trap. On the other hand, this peculiarity gives at the same time a strong analogy with biological cells, as cells are not equilibrium systems – and cannot be if they have to preserve their own identity (by mixing together horse liver cells and E. coli cells no average system is obtained). Thus, in the increase of self-organization that goes from micelles upwards, once the level of the double layer of vesicles is reached, the physical characteristics of life are already encountered (membranes that cannot be easily permeated by external solutes and refuse to comply with the laws of chemical equilibrium).

It is likely that early cells were more permissive, and perhaps an early step in the transition to life is the transition from permeable, simple protocells, to hard and impermeable structures, like our present POPC liposomes. In fact, the common stand of chemists to work with pure compounds may not be the best to model prebiotic systems. In a prebiotic scenario, most probably, mixtures of several surfactants and co surfactants were dominating the scene. It is known that the permeability of vesicles increases when co surfactants – like long-chain alcohols – are added. This observation about the importance of mixtures would in principle open the way to a vast area of research (see Sidebox 7.1).

In the meantime, the intense study of the simpler vesicle systems has unravelled novel, unsuspected physicochemical aspects – for example growth, fusion and fission, the matrix effect, self-reproduction, the effect of osmotic pressure, competition, encapsulation of enzymes, and complex biochemical reactions, as will be seen in the next chapter. Of course the fact that vesicles are viewed under the perspective of biological cell models renders these findings of great interest. In particular, one tends immediately to ask the question, whether and to what extent they might be relevant for the origin of life and the development of the early cells. In fact, the basic studies outlined in this chapter can be seen as the prelude to the use of vesicles as cell models, an aspect that we will considered in more detail in the next chapter.

Questions for the reader

1. The fusion of vesicles with opposite charges appears to be an efficient method to increase the molecular complexity. Suppose all enzymes of the Krebs cycle were distributed into five or six different charged vesicle families (some positively, some negatively charged) – then mixed. Would the Krebs cycle be reconstituted?
2. Vesicles can self-reproduce, however it is argued that there is no information content passing from one generation to the next. What do you think: is information really so important for self-reproduction of early protocells?
3. We have seen experiments of growth of vesicles, whereby fresh surfactant binds to the surface of pre-formed vesicles. How would you devise an experiment, so that the growth rate is determined – or affected – by the vesicle content?

11

Approaches to the minimal cell

Introduction

We come back now to the origin of life and the evolution of early cells. We have seen in the first chapters of this book the endeavor of people working with the aim of clarifying the pathway to the transition to life starting from simple molecules. This is the so-called bottom-up approach, the narrative by which a continuous and spontaneous increase of molecular complexity has transformed inanimate matter into the first self-reproducing entities, and from those, life at large.

The bottle neck of the bottom-up approach is the difficulty of reproducing on paper and/or in the laboratory those processes which have been moulded by contingency – such as the synthesis of specific macromolecular sequences.

There is another approach to the construction of the living cell, as indicated in Figure 11.1. This is to utilize the extant nucleic acids and enzymes and insert them into a vesicle, and re-construct in this way a minimal living cell.

Whereas the term "bottom-up" is recognized and accepted, the terminology of this alternative route to the minimal cell is less clear. The term "top-down" has been utilized by my group, mostly to set up a discrimination with respect to the bottom-up approach; however such a terminology is not really correct. In fact, this is also a bottom-up approach, in the sense that it goes in the direction of increasing complexity (the cell) starting from the single components.

The term "re-construction" comes to mind. However, by the procedure of Figure 11.1 the re-construction of an extant cell or something that exists on our Earth is not necessarily reached. The term "synthesis" rather than re-construction is better, synthesis in the sense of "synthetic biology." As a matter of fact, this is indeed a synthetic procedure, and since we are utilizing macromolecules that already exist, the best terminology is then the semi-synthetic approach, and the product will be a "semi-synthetic cell." The general term "artificial cells" has been used (Pohorille and Deamer, 2002). However, since generally extant enzymes/genes are used for

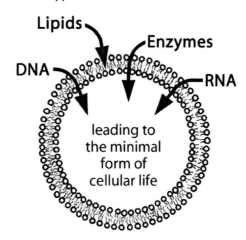

Figure 11.1 The semi-synthetic approach to the construction of the minimal cell.

those constructs, the term "semi-artificial cells" or "semi-synthetic cells" might be more appropriate.

The point of this procedure is not to synthesize a fully fledged modern cell, but the simplest possible form of it. To clarify this point, the notion of the "minimal cell" needs to be discussed is more detail.

The notion of the minimal cell

One of the earliest attempts to describe the DNA/proteins minimal cell was by Morowitz (1967). Based on the enzymatic components of primary metabolism, Morowitz estimated that the size of a minimal cell should be about one-tenth smaller than mycoplasma. There were earlier significant insights to the field by Dyson (1982) as well as by Woese (1983) and Jay and Gilbert (1987). More recently, the reviews by Deamer and coworkers (Pohorille and Deamer, 2002) together with other researchers (Ono and Ikegami, 2000; Luisi, 2002a; Oberholzer and Luisi, 2002) have sharpened the question and brought it into the perspective of the modern molecular tools.

Mycoplasma genitalium and *Buchnera* are considered the simplest cells, with a genome containing less than 500 coding regions. These are, however, parasites and the next step of complexity concerns microbes with thousands of expressed proteins, which catalyze thousands of reactions more or less simultaneously within the same tiny compartment – a maze of enormous complexity. However, precisely this complexity elicits the question of whether such complexity is really essential for life – or whether instead cellular life might be possible with a much smaller number of components.

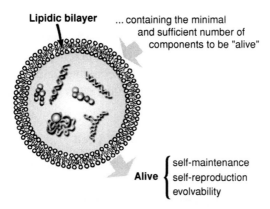

Lipidic bilayer ... containing the minimal
and sufficient number of
components to be "alive"

Alive { self-maintenance
self-reproduction
evolvability

Figure 11.2 The notion of minimal the cell. As explained in the text, this definition does not identify one particular structure, but is rather a descriptive term for a wide family.

This question is also implicit when considering the origin of life and early cells. Early cells could not have been as complex as our modern ones. Their enormous complexity is most likely the result of billions of years of evolution – with the development of a series of defense, repair, and security mechanisms, and also redundancies, and metabolic loops that are presently no longer essential. Thus the general question of how much the structure of modern cells can be simplified, is related to the question of the structure of the early cells.

This brings us to the notion of the minimal cell, defined as that having the minimal and sufficient number of components to be called alive. What does "alive" mean? Well, here we should go back to Chapter 2 and the various definitions of life; however, a fairly general definition can be used here, which ought to keep everybody satisfied: living at the cellular level means the concomitance of three properties: self-maintenance (metabolism), self-reproduction, and evolvability (see Figure 11.2).

If all these three properties are fulfilled, we have fully fledged cellular life. Of course in semi-synthetic systems the implementation can be less than perfect, and then several kinds of approximation to cellular life can be envisaged. For example, we can have protocells capable of self-maintenance but deprived of self-reproduction; or vice versa. Or we can have protocells in which self-reproduction is active for only a few generations; or systems that do not have the capability to evolve. Even in a given type of minimal cell – for example one with all three attributes – there might be quite different ways of implementation.

It is clear then that the term "minimal cell" depicts large families of possibilities. The question is how the minimal cells can be constructed in the laboratory. From the operational point of view, the illustrations of Figures 11.1 and 11.2 already suggest that the inclusion of components in lipid vesicles (liposomes) is the most obvious way to start. This chapter aims to provide some basic information on this, limited to experimental approaches, a choice that implies neglecting the many theoretical models of minimal life provided by computer scientists and theoreticians of complexity; and gliding over the experimental work made with cellular extracts in vitro.

The minimal RNA cell

Implicitly, this book has been concerned with biological cells as we know them on Earth, namely entities constituted by DNA, RNA, and proteins. However the simplest possible cell that on paper responds to the three criteria given in Figure 11.2 is an RNA cell. This theoretical object was described a few years ago (Szostak *et al.*, 2001) and in fact represents, among other things, a nice example of the coming together of the RNA world and the compartment world.

This is illustrated in Figure 11.3. It consists of a vesicle containing two ribozymes, one (Rib-2) capable of catalyzing the synthesis of the membrane component; the other (Rib1) being an RNA replicase that is capable of replicating itself, and reproducing the Rib-2 as well. In this way, there is a concerted shell-and-core replication, and there is therefore a basic metabolism, self-reproduction, and – since the replication mechanism is based on RNA replication – also evolvability.

How realistic is all this? One may recall at this point the previously mentioned self-replicating ribozyme reported by Paul and Joyce (2002) – however, we are still far from the production of an RNA polymerase that catalyzes the synthesis of itself as well as the synthesis of another ribozyme.

It is then a theoretical construction, and in addition the kind of scheme shown in Figure 11.3 implies a series of assumptions that we should be aware of. For example, it is assumed that both the precursor A and all mononucleotides, present in large excess outside, can permeate into the cell. It is also assumed that the cell divides by itself giving rise to identical daughters, whereas a statistical distribution of the macromolecules in the resulting cells should be expected (some having no Rib-1 and/or Rib-2; and only some having both of these ribosomes).

The construct of Figure 11.3 is still an ideal, hypothetical system, nevertheless very interesting in one respect: it shows that at least in principle, cellular life can be implemented by a very limited number of RNA genes. Those who believe in the RNA world may add that this basic simplicity indicates the predominant importance of RNA in the early stages of life.

The RNA cells eventually have to evolve into protein/DNA cells. And this is a long and certainly not easy pathway. However, this is the beauty of the "prebiotic" RNA world: that at least on paper, a possible pathway leading to DNA and proteins can be conceived. One ideal pathway showing the transition from the RNA to the DNA cell is illustrated in the Figure 11.4.

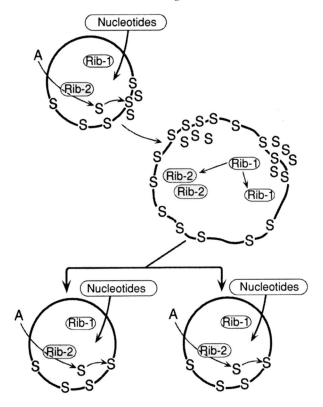

Figure 11.3 The simplest RNA cell, consisting of two ribozymes (two RNA-genes provided with enzymatic activity), Rib-1(ribosome 1) and Rib-2 (ribosome 2), whose concerted action permits shell and core replication. Rib-1 is an RNA replicase, capable of making copies of itself and of Rib-2. Rib-2 makes the lipid membrane, converting precursor A to surfactant S. Being based on RNA replication, it is also able to evolve. (Adapted from Szostak *et al.*, 2001; see also Luisi *et al.*, 2002.)

The minimal genome

Let us now go back to the more familiar DNA world with the question of the minimal genome. This has been considered by several authors, for example, Mushegian and Koonin (1996); Shimkets (1998); Mushegian (1999); Koonin (2000); Kolisny-chenko *et al.* (2002); Luisi *et al.* (2002); Gil *et al.* (2004); Islas *et al.* (2004). Table 11.1 gives an overview of some salient contributions in the field.

Mushegian and Koonin (1996) calculated an inventory of 256 genes that represents the amount of DNA required to sustain a modern type of minimal cell under permissible conditions. This number, as indicated later by Koonin (2000), is quite similar to the values of viable minimal genome sizes inferred by site-directed gene disruptions in *B. subtilis* (Itaya, 1995) and transposon-mediated

Table 11.1. *Some miniature cellular genomes (from Islas et al., 2004)*

Species	Genome size (kb)	Lifestyle	Reference
Mycoplasma genitalium	580	Obligate parasite	Fraser *et al.*, 1995
Buchnera spp.	450	Endosymbiont	Gil *et al.*, 2002
Crytomonad nucleomorph	551	Secondary endosymbiont	Douglas *et al.*, 2001

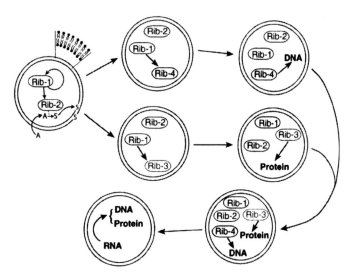

Figure 11.4 The hypothetical pathway for the transformation of a simple RNA cell into a minimal DNA/protein cell. At the first step, the cell contains two ribozymes, Rib-1 and Rib-2: Rib-1 is a RNA replicase capable of reproducing itself and making copies of Rib-2, a ribozyme capable of synthesizing the cell membrane by converting precursor A to surfactant S. During replication, Rib-1 is capable of evolving into novel ribozymes that make the peptide bond (Rib-3) or DNA (Rib-4). In this illustration, these two mutations are assumed to take place in different compartments, which then fuse with each other to yield a protein/DNA minimal cell. Of course, a scheme can be proposed in which both Rib-3 and Rib-4 are generated in the same compartment. (Modified from Luisi *et al.*, 2002.)

mutagenesis knock-outs in *M. genitalium and M. pneumonia* (Hutchinson *et al.*, 1999).

On the subject of this last work, one may recall that the notion of "minimal genome" is approached in quite a different way by Craig Venter and his group. In work carried out at the Institute for Genomic Research in Rockville, Maryland, a team led by Clyde Hutchinson knocked out genes from a *M. genitalium* bacterium

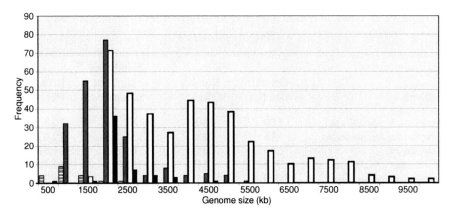

Figure 11.5 Prokaryotic genome size distribution ($N = 641$). Open boxes, free-living prokaryotes; grey boxes, obligate parasites; black boxes, thermophiles; boxes with horizontal lines, endosymbionts. (Modified from Islas *et al.*, 2004.)

one by one, and estimated that out of the 480 protein-coding regions, about 265 to 350 are essential under laboratory growth conditions, including about 100 genes with unknown functions (Hutchison *et al.*, 1999; Mushegian, 1999).

More recently, in collaboration with Hamilton Smith, Craig Venter attempted something more ambitious, namely the construction of a minimal mycoplasma genome by building its chromosome from scratch with the use of chemical synthesizers (Smith *et al.*, 2003). The authors described how they dramatically shortened the time required for accurate assembly of genomic material, for example, they could assemble in 14 days the complete infectious genome of a 5386 bp bacteriophage from a single pool of chemically synthesized oligonucleotides. This approach was used by Eckard Wimmer at Stony Brook to create an infectious poliovirus, which is much simpler than a bacterium (Cello *et al.*, 2002).

Going one step further, Venter's idea was to remove the original genetic material from the bacterium and insert the synthetic one to see whether it works (Zimmer, 2003). Most biologists would agree that there is no reason why it should not work. The vitalistic idea that there is something special in DNA has long since gone from present-day scientific thinking. We all accept the idea that DNA is just a molecule like any other, and therefore there will be no great surprise if Venters' synthesis of the bacterium genome is crowned by success. However, his courage and determination to fulfil such a formidable synthetic and organizational enterprise is still very admirable.

Let us now come back to nature as it is, and look at the size of the smallest organisms. Figure 11.5 compares the genome size distribution, calculated under a series of assumptions (Islas *et al.*, 2004), of free-living prokaryotes, obligate parasites, thermophiles, and endosymbionts. The values of DNA content of free-living

Approaches to the minimal cell

Table 11.2. *Genetic redundancies in small genomes of endosymbionts and Obligate parasites*

Proteome	Genome size (kb)	Number of ORFs	Number of redundant sequences	Redundancy (%)
Mycoplasma genitalium	580	480	52	10.83
Mycoplasma pneumoniae	816	688	134	19.47
Buchnera sp.	640	574	67	11.67
Ureaplasma urealyticum	751	611	105	17.18
Chlamydia trachomatis	1000	895	60	6.71
Chlamydia muridarum	1000	920	60	6.52
Chlamydophila pneumoniae	1200	1070	148	13.83
Rickettsia prowazekii	1100	834	49	5.87
Rickettsia conorii	1200	1366	189	13.83
Treponema pallidum	1100	1031	78	7.56

* Genome sizes, complete proteomes, and the number of open reading frames (ORFs) were all retrieved from http://www.ncbi.nlm.nih.gov. Taken, with permission, from Islas *et al.*, 2004.

prokaryotes can vary over a tenfold range, from 1450 kb for *Halomonas halmophila* to the 9700 kb genome of *Azospirillium lipoferum* sp59b. By way of comparison, consider that *Escherichia coli K-12* has a genome size of *c.* 4640 kb and *Bacillus subtilis* 4200 kb.

Classification of endosymbionts as a group by themselves shows that their DNA content can be significantly smaller, reaching a value of 450 kb for *Buchnera* spp. (Gil *et al.*, 2002) or a value of 580 kb for the obligate parasite *Mycoplasma genitalium* (Fraser *et al.*, 1995).

The smallest sizes are then those of *Mycoplasma genitalium* and *Buchnera*, with a value that agrees well with the predictions of Shimkets (1998), according to which the minimum genome size for a living organism should be approximately 600 kb. It is argued that these two organisms have undergone massive gene losses and that their limited encoding capacities are due to their adaptation to the highly permissive intracellular environments provided by the hosts (Islas *et al.*, 2004).

What do these numbers mean in terms of minimal number of genes? Table 11.2, taken from Islas *et al.* (2004), reports the number of coding regions in some small genomes. The table also gives an account of the redundant genes, amounting on average to 6–20% of the entire genome. Redundant genes stem from paralogous genes, which are sequences that diverge not throughout speciation but after a duplication event. Based on these and other data, the authors also provide indications for the existence of a more primitive, less regulated version of protein synthesis. In fact, data from in vitro translation systems support the possibility of an older

Table 11.3. *Core of a minimal bacterial gene set**

	Number of genes
DNA metabolism	**16**
Basic replication machinery	13
DNA repair, restriction, and modification	3
RNA metabolism	**106**
Basic transcription machinery	8
Translation: aminoacyl-t-RNA synthesis	21
Translation: t-RNA maturation and modification	6
Translation: ribosomal proteins	50
Translation: ribosome function, maturation and modification	7
Translation factors	12
RNA degradation	2
Protein processing, folding and secretion	**15**
Protein post-translational modification	2
Protein folding	5
Protein translocation and secretion	5
Protein turnover	3
Cellular processes	**5**
Energetic and intermediary metabolism	**56**
Poorly characterized	**8**
Total	**206**

* Courtesy of Andres Moya, Institut Cavanilles de Biodiversitat i Biologia Evolutiva, Universitat de València (España).

ancestral-protein synthetic apparatus prior to the emergence of elongation factors (Gavrilova *et al.*, 1976; Spirin, 1986).

How can the data in Table 11.2 be used in order to envisage further possible simplification of the genome? Andres Moya and his group in Valencia arrived at the smaller number of 206 genes on the basis of their work with *Buchnera sp.* and other organisms (Gil *et al.*, 2004). The results are given in Table 11.3, kindly provided by Andres Moya. The number of 206 genes as minimal genome represents on one hand a considerable simplification. On the other hand, it still corresponds to a formidable complexity giving rise to the question of just how much further down it is possible to go.

Further speculations on the minimal genome

Obviously, only speculations can help at this point. One way to speculate is to imagine a kind of theoretical knock down of the genome, reducing cellular complexity and at the same time part of the non-essential functionality (Luisi *et al.*, 2002).

The first step in this intellectual game is to imagine a cell without the genes (the enzymes) needed to synthesize low-molecular-weight compounds, assuming that low-molecular-weight compounds, including nucleotides and amino acids, are available in the surrounding medium and able to permeate into the cell membrane. This would be a "fully permeable minimal cell." A high permeability is in principle possible, as shown recently by Noireaux and Libchaber (2004), who described how by the presence of α-hemolysine on the cell membrane, low-molecular substrates are imported inside a vesicle. Further simplification (Luisi *et al.*, 2002) finally gives a cell that would be able to perform protein and lipid biosynthesis by a modern ribosomal system, but limited to a restricted number of enzymes – see Table 11.4. This cell would have *c.* 25 genes for the entire DNA/RNA synthetic machinery, *c.* 120 genes for the entire protein synthesis (including RNA synthesis and the 55 ribosomal proteins), and 4 genes for the synthesis of the membrane. This would bring to a total of *c.* 150 genes (first column of Table 11.4), somewhat less than the figure of 206 exposed previously.

Thanks to the outside supply of substrates, such a cell should be capable of self-maintenance and of self-reproduction, including replication of the membrane's components. However it would not make low-molecular-weight compounds and would not have redundancies for its own defense and security – all self-repair mechanisms are missing. Also, there is the problem of leakage and lack of concentration gradients; and cell division would be simply due to a physically based statistical process.

There is of course no proof that this theoretical construct would work – and this also holds of course for Moya's 206 genes. It is nevertheless instructive to go further with these theoretical knock down experiments.

The next victims would be ribosomal proteins. Can we take them out? There are some indications that ribosomal proteins may not be essential for protein synthesis (Zhang and Cech, 1998), and there are other suggestions about an ancient and simpler translation system (Nissen *et al.*, 2000; Calderone and Liu, 2004). If we accept this, and take out the 55 genes of the ribosomal proteins and some other enzymes, around 110 genes (second column of Table 11.4) would be obtained.

Can we make further reductions?

A large portion of the above mentioned genes correspond to RNA and DNA polymerases. A number of data (Suttle and Ravel, 1974; Lazcano *et al.*, 1988; 1992; Frick and Richardson, 2001) suggest that a simplified replicating enzymatic repertoire – as well as a simplified version of protein synthesis – might be possible. From all this, the idea that a single polymerase could play multiple roles as a DNA polymerase, a transcriptase, and a primase, is conceivable in the very early cells (Luisi *et al.*, 2002).

The game could go on by assuming that at the time of the early cells not all 20 amino acids were involved – and a lower number of amino acids would

Table 11.4. *A list of genes that define minimal cells, sorted by functional category*

	Number of genes		
Gene function	Minimal DNA cell[a]	"Simple-ribosome" cell	Extremely reduced cell
DNA/RNA metabolism			
DNA polymerase III	4[b]	4[b]	1
DNA-dependent RNA polymerase	3[c]	3[c]	1
DNA primase	1	1	
DNA ligase	1	1	1
Helicases	2–3	2–3	1
DNA gyrase	2[d]	2[d]	1
Single-stranded-DNA-binding protein	1	1	1
Chromosomal replication initiator	1	1	
DNA topoisomerase I and IV	1 + 2[d]	1 + 2[d]	1
ATP-dependent RNA helicase	1	1	
Transcript. elongation factor	1	1	
RNAases (III, P)	2	2	
DNAases (endo/exo)	1	1	
Ribonucleotide reductase	1	1	1
Protein biosynthesis/translational apparatus			
Ribosomal proteins	51	0	0
Ribosomal RNAs	1 operon with 3 functions (r-RNAs)	1 operon with 3 functions (r-RNAs)	1 operon with 3 functions (r-RNAs), self splicing
Aminoacyl-t-RNA synthetases	24	24	14[e]
Protein factors required for protein biosynthesis and synthesis of membrane proteins	9–12[f]	9–12[f]	3
Transfer RNAs	33	33	16[g]
Lipid metabolism			
Acyltransferase "pIsX"	1	1	1
Acyltransferase "pIsC"	1	1	1
PG synthase	1	1	1
Acyl carrier protein	1	1	1
Total	**146–150**	**105–107**	**46**

[a] Based on *M. genitalium*. [b] Subunits *a, b, y,* tau. [c] Subunits *a, b, b'.* [d] Subunits *a, b*.
[e] Assuming a reduced code. [f] Including the possible limited potential to synthesize membrane proteins. [g] Assuming the third base to be irrelevant.

reduce the number of amino acyl-t-RNA-synthetases and the number of t-RNA genes.

All these considerations may help to decrease the number of genes down to a happy number of, say, 45–50 genes – see last column of Table 11.4 – for a living, although certainly limping, minimal cell (Luisi *et al.*, 2002).

This number is significantly lower than that proposed by Moya in the previous table, but is of course based on a higher degree of speculation. Many authors would doubt that a cell with only 45–50 genes would be able to work. However, again, the consideration moves to early cells, and to the argument that the first cells could have not started with hundreds of genes from the very beginning in the same compartment. In fact, there are some claims that the first ribosomes consisted of r-RNAs associated simply with basic peptides (Weiner and Maizels, 1987).

The consideration of the minimal cell permits a logical link with the notion of compartments outlined in the previous chapter. Suppose that these 45–50 macro-molecules – or their precursors – developed first in solution. In order to start cel-lular life, compartmentation should have come later on, and one would then have to assume the simultaneous entrapment of all these different genes in the same vesicle. This can indeed be regarded as highly improbable. A more reasonable sce-nario is one in which the complexity of cellular life evolved from within the same compartment – a situation namely where the 45 (or 206) macromolecules were produced and evolved from a much smaller group of components from the inside of the protocell. How, of course, remains to be seen.

The road map to the minimal cell. 1: Complex biochemical reactions in vesicles

Having outlined what the minimal chimerical genome may look like, we can tackle the question of how the construction of minimal cells can be approached in the laboratory.

Such a research project is a complex enterprise and it may be useful to divide up the "road map" to the minimal cell into different milestones of increasing com-plexity. The first one, which is already under control in several laboratories, is to carry out and optimize complex enzymatic reactions in liposomes – such as the polymerase chain reaction, the biosynthesis of RNA and DNA, the condensation of amino acids, etc.

Perhaps one of the very first examples of enzymatic reactions carried out in liposomes with the aim of building a minimal cell is the work by Schmidli *et al.* (1991), as already mentioned in the previous chapter (Fig. 10.5). The general idea is illustrated in Figure 11.6, whereas the biochemical pathway is illustrated in Figure 11.7. The basic idea is to have inside the liposomes the series of reac-tions that, starting from a relatively simple product (G3P, glycerol-3-phosphate)

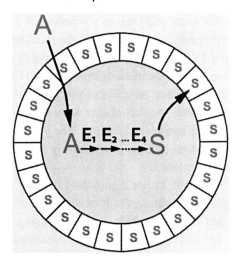

Figure 11.6 A liposome that builds its own membrane with the help of entrapped enzymes: the prototype of the simplest autopoietic minimal cell. In the experiment by Schmidli *et al.* (1991), four different enzymes were entrapped in one single lipsome, with the idea of synthesizing lecithin (see the reaction scheme in Figure 11.7). See also Fig. 10.5.

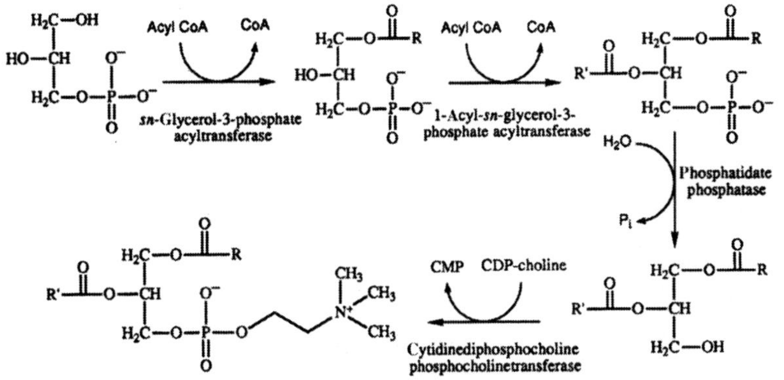

Figure 11.7 Salvage pathway for phosphatidylcholine synthesis. (Modified from Schmidli *et al.*, 1991.)

leads to the phospholipid membrane. In Figure 11.7 the Salvage pathway for phosphatidylcholine synthesis can be seen, with the four enzymes G3P-AT (*sn*-glycerol-3-phosphate acyltransferase), LPA-AT (1-acyl-*sn*-glycerol-3-phosphate acyltransferase), PA-P (phosphatidate phosphatase), and CDPC-PT (cytidinediphosphocholine phosphocholinetransferase). It was shown, as expected, that these enzymes were active when associated with liposomes, with an activity corresponding to that in the original microsomes. It was shown that phosphatidylcholine (PC) was synthesized in the enzyme-containing liposomes, and that its yield was about the same when analyzed as a function of the radioactive *sn*-G3P as substrate or when cytidinediphosphocholine is the radioactive-labeled substrate. The other substrates used for the reaction were palmitoyl and oleoyl coenzyme A (CoA), so that palmitoyloleylphosphatidylchsline (POPC) was the synthetized product. In order to visualize better the effect of this synthesis on the physical state of the proteoliposomes, shorter lecithin chains were also synthesized using hexanoyl CoA and dihexanoyl phosphatidylcholine. The yield was only around 10%, and the effect on the liposome size, as examined by light scattering, was in the expected direction, but rather small.

This kind of work, although rather significant, has not been repeated, partly because of the difficulty of obtaining the four enzymes with a sufficient degree of purity. Some work has been done with the idea of using liposomes formed by phosphatidic acid (the product of the first two enzymes shown in Figure 11.7), see Luci (2003).

A different example of enzymatic synthesis in vesicles was also mentioned in the previous chapter : the polycondensation of adenonine diphosphate (ADP) into poly(A) by encapsulated enzymes (Chakrabarti *et al.*, 1994; Walde *et al.*, 1994a). In the case of Walde *et al.* (1994a) there was a core-and-shell reproduction, as the synthesis of poly(A) – a prototype of RNA – was occurring simultaneously with the self-reproduction of the vesicle shells (see Figure 10.3).

A suggestive example of core and shell reproduction was provided shortly after this (Oberholzer *et al.*, 1995b) with the use of the famous Spiegelmann and Eigen's enzyme, the Qβ replicase. As illustrated in Figure 11.8, while the enzyme was replicating RNA, the oleate vesicles were multiplying on their own accord. The vesicle self-reproduction was induced by the binding of the water-insoluble oleic acid anhydride, as described earlier. The hydrolysis of the anhydride was followed spectroscopically by FTIR (Fourier transform infrared), and the kinetics of the reaction, as well as the vesicle size distribution, were studied by freeze-fracture electron microscopy. The vesicles, in addition to the enzyme and RNA MDV-1 template, contained the triphosphates of adenine, cytosine, guanine and uracil, i.e. ATP, CTP, GTP, UTP, (^{35}S ATP), as well as Mg^{2+} ions and buffer. This experiment was operated under excess of Qβ replicase/RNA template, so that the replication of RNA could proceed for a few generations.

At first sight, Figure 11.8 already appears to correspond to a living cell. Still, since reproduction of vesicles occurs statistically, after a while the large majority of new vesicles will not contain either enzyme or template, and therefore the system will undergo a "death by dilution." Death by dilution is typical of all cell systems where there is no regeneration of the macromolecular components. A further shortcoming of the system is given by the fact that the two self-reproduction processes are not coupled with each other.

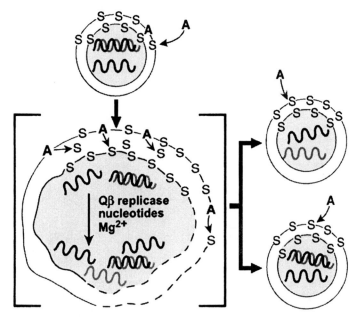

Figure 11.8 Replication of RNA in self-reproducing vesicles. The initial vesicles contained the enzyme Qβ replicase and the four ribonucleotides in excess, as well as the RNA template (the MDV-1 template). The division of vesicles is induced by the addition of oleic acid anhydride and the duplication of the figure is idealized, as in reality division occurs on a statistical basis. (Adapted from Oberholzer *et al.*, 1995b.)

Another complex biochemical reaction is the "polymerase chain reaction" (PCR). It has also be implemented in liposomes (Oberholzer *et al.*, 1995a). This reaction was interesting from the point of view of vesicle chemistry because the liposomal system had to endure the extreme PCR conditions, with several temperature cycles up to 90 °C (liposomes were practically unchanged at the end of the reaction) and furthermore, nine different chemicals had to be encapsulated in an individual liposome for the reaction to occur. This was carried out by mechanical entrapment from a solution that contained all components: only a minimal number of the *in situ* formed vesicles could entrap all nine components. These odds notwithstanding, there was a significant synthesis.[1]

---

[1] In fact if one calculates the probability that all nine components are trapped by chance inside one single vesicle, this is extremely small, and of course becomes smaller, the larger the number of components. A few such experiments have been described and there is an interesting point in this regard, which has never been studied in detail: it appears that the number of highly filled vesicles is generally higher than that expected on a statistical basis – as if there were some cooperativity effect that facilitates the uptake of components, once a certain number of components is already inside. This phenomenon was apparent, for example, in the incorporation of ferritin molecules in oleate vesicles, where liposomes were found (by TEM) containing up to 30–50 ferritin molecules per vesicle (Berclaz *et al.*, 2001a).

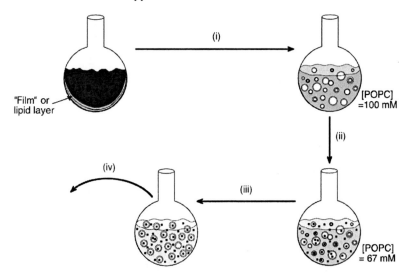

Figure 11.9 Schematic view of the experimental strategy for carrying out poly(Phe) synthesis in POPC liposomes. (i) Freeze–thaw ($\times 7$); solution containing t-RNA^{phe}, poly(U), [^{14}C] Phe, ATP, GTP, Mg(OAc)$_2$, NH$_4$Cl, spermine, spermidine, phosphoenolpyruvate. (ii) Soution containing pyruvate kinase, 100 000 g supernatant enzymes, 30S and 50S ribosomal subunits. (iii) 1. Free-thaw ($\times 3$); 2. Brief extrusion; 3. Addition of EDTA (final concentration = 35 mM. (iv) Withdrawl of aliquots at indicated time and cold TCA precipitation. Analysis of the radioactivity remaining on the glass filter by β-scintillation counting. (Modified from Oberholzer *et al.*, 1999.)

A significant step forward in this field has been provided by the entrapment of an entire ribosomal system in POPC vesicles (Oberholzer *et al.*, 1999). In this first experiment, only poly(Phe) was synthesized, using poly(U) as the m-RNA. Although this work is now obsolete, it contains several elements of interest. Firstly, several components must again be entrapped within the same liposome in order for the polypeptide synthesis to occur: the entire ribosomal structure, namely the complex between the subunit 50S and 30S; the messenger RNA (poly (U) in this case), t-RNA^{phe}, the elongation factors EF-Tu, EF-G, EF-Ts, and the substrate Phe (labeled in this case). The strategy of choice was again to form the liposomes in a solution containing all components, and the entrapment yield was increased by way of several freeze and thaw cycles. All procedures for the preparation of liposomes had to be carried out at 5 °C in order to avoid poly(Phe) synthesis outside the liposomes; furthermore, after the extrusion step, ethylenediaminetriacetic acid (EDTA) in excess was added to the external solution in order to inhibit the poly(Phe) expression outside the liposomes. The whole procedure is illustrated in Figure 11.9. The yield was 5% with respect to the experiment in water without liposomes, and the authors argue (Oberholzer *et al.*, 1999) that this yield is actually surprisingly high, considering that the liposomes occupy only a very small fraction of the total volume of solution; and that only a very few of them would contain all ingredients by the statistical entrapment.

In the years immediately following, this procedure was optimized to express real proteins, the first being – for obvious detection reasons – the green fluorescence protein (GFP). Protein expression is actually the subject of the next section.

The road map to the minimal cell. 2: Protein expression in vesicles

What can be done in order to approach the construction of the minimal cell? In principle, the complexity of the core of the liposomes should be increased so as to reach the minimal genome.

As already mentioned, this has not been the approach used until now in the literature. Rather, people have first sought to insert into liposomes the conditions to express a single protein. For reasons easy to understand (detection facility), the green fluorescence protein (GFP) has been the target protein.

With how many genes? Well, this question is also not easily answered from current data, as generally the authors have not calculated the number of genes/enzymes involved. Often, then, commercial kits are used for protein expression, and these kits are notoriously black boxes where the number of genes/enzymes is not made known. Occasionally, the entire *E. coli* cellular extract has been utilized, although for the expression of one single protein only a minimal part of the *E. coli* genome will be utilized.

An overview of the work in this field is presented in Table 11.5, see also recent reviews (Luisi *et al.*, 2006). This table also contains references to the work mentioned earlier, such as poly(A) synthesis from ADP; the PCR reaction in liposomes; the RNA synthesis by Qβ replicase, as well as the expression of poly(Phe) by an entrapped ribosomal system. This work is preliminary to protein expression in liposomes. Going from here to the protein synthesis, it may be useful to compare the different strategies for the expression of GFP.

The common strategy is to entrap into the aqueous core of liposomes all the ingredients for the in vitro protein expression; i.e., the gene for the GFP (a plasmid), an RNA polymerase, ribosomes, and all the low-molecular-weight components (amino acids, ATP, etc.) needed for protein expression.

Yomo, Urabe and coworkers (Yu *et al.*, 2001), for example, reported the expression of a mutant GFP (actually the pET-21-GFPmutl-His6 mutant) in lecithin liposomes. Large GFP-expressing vesicles, prepared by the film hydration method, were analyzed using flow cytometry as well as confocal laser microscopy.

In the procedure utilized by Oberholzer and Luisi (2002) all ingredients are added to a solution in which the vesicles are being formed by the ethanol injection method,

Table 11.5. *Synthesis of nucleic acids and protein expression in vesicles*

Description of the system	Main goal and results	References
Enzymatic poly(A) synthesis.	1. Polynucleotides phosphorilase producing poly(A) from ADP; 2. poly(A) is produced inside simultaneously with the (uncoupled) self-reproduction of vesicles.	Chakrabarti et al., 1994. Walde et al., 1994.
Oleic acid/oleate vesicles containing the enzyme Qβ replicase, the RNA template and the ribonucleotides. The water-insoluble oleic anhydride was added externally.	A first approach to a synthetic minimal cell: the replication of a RNA template proceeded simultaneously with the self-replication of the vesicles.	Oberholzer et al., 1995b.
POPC liposomes containing all different reagents necessary to carry out a PCR reaction.	DNA amplification by the PCR inside the liposomes; a significant amount of DNA was produced.	Oberholzer et al., 1995a.
POPC liposomes incorporating the ribosomal complex together with the other components necessary for protein expression.	Ribosomal synthesis of polypeptides can be carried out in liposomes; synthesis of poly(Phe) was monitored by TCA of the ^{14}C-labelled products.	Oberholzer et al., 1999.
Liposomes from EggPC, cholesterol, DSPE-PEG5000 used to entrap cell-free protein synthesis.	Expression of a mutant GFP, determined with flow cytometric analysis.	Yu et al., 2001.
T7RNA polymerase within cell-sized giant vesicles formed by natural swelling of phospholipid films	Transcription of DNA and transportation by laser tweezer; vesicles behaved as a barrier preventing the attack of RNAase.	Tsumoto et al., 2002.
DNA template and the enzyme T7RNA polymerase microinjected into a selected giant vesicle; nucleotide triphosphates added from the external medium	The permeability of giant vesicles increased in an alternating electric field; m-RNA synthesis occurred.	Fischer et al., 2002.

Reaction initiated by mixing all reagents in water, then transferring to POPC liposomes.	Expression of GFP in liposomes.	Oberholzer and Luisi, 2002
Gene-expression system within cell-sized lipid vesicles.	Encapsulation of a gene-expression system; expression of GFP with very high efficiency.	Nomura *et al.*, 2003.
A water-in-oil compartment system with water bubbles up to 50 μm.	Expression of GFP by mixing different compartments which are able to fuse with each other.	Pietrini and Luisi, 2004.
E. coli cell-free expression system encapsulated in a phospholipid vesicle, which was transferred into a feeding solution containing ribonucleotides and amino acids.	The expression of the α-hemolysin pore protein from *S. aureus* inside the vesicle solved the energy and material limitations; the reactor could sustain expression for up to four days.	Noireaux *et al.*, 2004.
A two-stage genetic network encapsulated in liposomes.	A genetic network in which the protein product of the first stage (T7RNA polymerase) is required to drive the protein synthesis of the second stage (GFP).	Ishikawa *et al.*, 2004.

and EGFP (enhanced GFP) production is then evidenced inside the compartments. In this case, the sample was analyzed spectroscopically, monitoring the increase of the fluorescent signal of the EGFP. The disadvantage of this procedure is that entrapping efficiency is generally low, due to the small internal volume of liposomes obtained with this method.

This problem is partly avoided in the procedure utilized by Yomo and Nomura's (Nomura *et al.*, 2003) by using giant vesicles. The reaction is observed by laser-scanning microscopy and shows that expression of rsGFP (red shifted GFP) takes place with a very high efficiency (the concentration of rsGFP inside vesicles was greater than that in the external environment). The authors also showed that vesicles can protect gene products from external proteinase K.

Based on the initial report on the expression of functional protein into liposomes by Yomo and coworkers (Yu *et al.*, 2001), the work by Ishikawa *et al.* (2004) represents another stage of the work on GFP expression. In fact, a two-stage genetic network is described, where the first stage is the production of T7 RNA polymerase, required to drive the GFP synthesis as the second stage.

Pietrini and Luisi (2004) described the synthesis of GFP from mixing two or more initial reagents, utilizing water in oil emulsions, see Figure 9.17. This system is biologically less interesting, however, compartments can fuse with each other and they exhibit no leakage.

Going back to vesicles, of particular interest is the work by Noireaux and Libchaber (2004). Again, a plasmid encoding for two proteins was used; in particular, the authors introduced EGFP and α-hemolysin genes. At variance with the cascading network described above, the second protein (α-hemolysin) does not have a direct role in protein expression, but is involved in a different task. In fact, although α-hemolysin is a water-soluble protein, it is able to self-assemble as a heptamer in the bilayer, generating a pore 1.4 nm in diameter (cut-off ∼3 kDa). In this way, it was possible to feed the inner aqueous core of the vesicles, realizing a long-lived bioreactor, where the expression of the reported EGFP was prolonged up to four days. This work represents an important milestone in the road map to the minimal cell, because the α-hemolysin pore permits the uptake of small metabolites from the external medium and thus solves the energy and material limitations typical of the impermeable liposomes, with the limitations mentioned earlier. Interesting is also the previous work by the same group, in which a cell-free genetic circuit assembly (without vesicles) is described (Noireax *et al.*, 2003). In particular, the authors engineered transcriptional activation and repression cascades, in which the protein product of each stage is the input required to drive or block the following stage. The expression of cascading genetic networking is studied also by Ishikawa *et al.* (2004) and most probably is the direction to consider for development of semi-artificial cell systems.

Table 11.5 reports also the work by Fischer *et al.* (2002) on m-RNA synthesis inside giant vesicles utilizing a DNA template and T7 RNA polymerase; and the transcription of DNA by Tsumoto *et al.* (2002).

In concluding this section, it is important to mention some interesting studies on microtubulation. Although at first sight not directly related to the question of the minimal cell, this kind of work paves the way for studying the intracellular transport in semi-artificial cells. The combination of giant vesicles, minibeads, and molecular motors has been studied by a team from the Institut Curie (Roux *et al.*, 2002). The authors showed that lipid giant unilamellar vesicles to which kinesin molecules were attached give rise to membrane tubes and to complex tubular networks that can form an original system to emulate intracellular transport. Membrane tube formation from giant vesicles by dynamic association of motor proteins has been studied by Koster *et al.* (2003), and in another context by Samkararaman *et al.* (2004). Also Tabony's group is active in this field (Glade *et al.*, 2004).

The road map to the minimal cell. What comes next?

Keeping in mind the notion of the minimal cell, the analysis of the data presented in Table 11.5 and in the discussion in the text makes clear what is still missing in order to proceed in the field. For example, protein expression, as outlined in the most salient experiments outlined in Table 11.5, has been carried out without checking the number of enzymes/genes utilized in the work. I believe that it would be proper to carry out protein expression utilizing known concentrations of the single enzymes/genes – so as to know what is in the pot, and possibly have a handle on the corresponding chemical equilibria. This operation would correspond to the implementation of the minimal genome inside liposomes and could pave the road to all the next steps.

Table 11.5 and the discussion in the text also clarify one other essential element that is still missing to reach the ideal minimal cell: self-reproduction. In fact, none of the systems described so far is able to reproduce itself after producing GFP.

In real biological systems, a cell is capable of duplicating and reproducing itself with the same genetic content. This is due to complex systems of regulation, which is not really compatible with the experimental set up of minimal cells. For the time being, some alternative approaches to the question of self-reproduction should be envisaged. (Luisi *et al.*, 2006).

One possibility is to achieve vesicle self-reproduction by the endogenous synthesis of the vesicle lipid building-blocks. Two strategies can be in principle pursued: (i) incorporating the enzymes that metabolize the lipids or, (ii) starting from the corresponding genes, i.e., expressing those enzymes within the vesicles (see Figure 11.10).

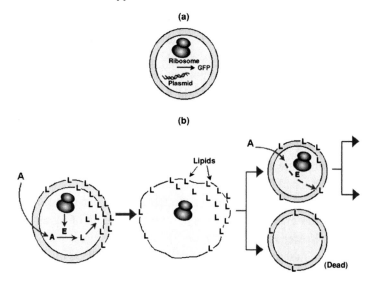

Figure 11.10 Protein expression inside the liposomes: a working plan. Schematic illustration of two critical steps on the road map to the minimal cell: (a) Protein expression of a simple protein (GFP) or any other simple protein and (b) protein expression of the enzymes that catalyze the formation of the vesicle boundary. For the sake of simplicity, growth and division is illustrated as an ideal duplication.

As already mentioned, early attempts have been focused on the enzymatic production of lecithin in lecithin liposomes (Schmidli *et al.*, 1991). The metabolic pathway was the so-called Salvage pathway, which converts glycerol-3-phosphate to phosphatidic acid, then diacylglycerol and finally phosphatidylcholine. Production of the cell boundary from within corresponds to autopoiesis and would close the circle between minimal cell and the autopoietic view of cellular life.

The internal synthesis of lecithin in lecithin liposomes would be a significant step forwards. In particular, it would be very interesting to see, given a certain excess of the enzymes, for how many generations the cell self-reproduction could go on. It is clear, however, that after a certain number of generations, the system would undergo "death by dilution."

Finally, in order to get closer to the real minimal cell, there is the problem of further reduction of the number of genes. In all the systems in Table 11.5, we are still dealing with ribosomal protein biosynthesis and this implies at least 100–200 genes. We are still far from the ideal picture of a minimal cell and can again pose the question of how to reduce this complexity.

As a way of thinking, we have to resort to the conceptual knock-down experiments, for examples those outlined in the work by Luisi *et al.* (2002) and Islas *et al.* (2004) – a simplification that also corresponds to a movement towards the early cell.

The simplification of the ribosomal machinery, and of the enzyme battery devoted to RNA and DNA synthesis, is probably a necessary step. (Luisi *et al.*, 2006).

Several questions can be posed at this level. For example: can simple matrices be developed that are operative in vitro as ribosomes? Can one operate, at costs of specificity, with only a very few polymerases? Similarly, reasoning that it might not be necessary for the very beginning to have all specific t-RNAs, can a few unspecific ones be used instead? Also, experiments with a limited number of amino acids might even be conceived.

Now, all this must be tried out experimentally, there is no way around this. This is where research on the minimal cell will have to be concentrated in the next few years. This leads on to the concluding remarks.

Concluding remarks

The definition of the minimal cell as given at the beginning of this chapter although simple has its own elegance. Conversely, the experimental implementations described so far may appear awkward or at least not as satisfactory and elegant. There are indeed objective difficulties still facing the construction of a minimal cell. We have seen for example that in the best case death by dilution is one limit we probably have to live with. Generally, the constructs realized in the laboratory until now represent still poor approximations of a fully fledged biological cell. The gap between this and real biological cells is such, that the possible bioethical hazards of the field of the minimal cell can for the moment be discounted.

Yet, just the conceiving and the study of these forms of "limping life," represent in my opinion the most interesting part of this on-going research. In fact, these approximations to life, such as a cell that produces proteins and does not self-reproduce; or one that does self-reproduce for a few generations and then dies out of dilution; or a cell that reproduces only parts of itself; and/or one characterized by a very poor specificity and a very poor metabolic rate . . . all these may and probably are intermediates experimented with by nature to arrive at the final destination, the fully-fledged biological cell. Thus, the realization in the laboratory of these partially living cells may be of fundamental importance to understand the real essence of cellular life, as well as the historical evolutionary pathway by which the final target may have been reached. It is true, however, that construction of a semi-synthetic cell by using extant enzymes and nucleic acids is not the solution to the origin of life. For that, we have to find ways by which such functional macromolecules are produced in a prebiotic world – and we have seen that this is not yet understood.

From a different perspective, the construction in the laboratory of a semi-synthetic living cell would be a demonstration – if still needed – that life is indeed an emergent property. In fact in this case cellular life would be created from non-life – as single genes and/or single enzymes are per se non-living.

All this is very challenging, and perhaps for this reason there has been an abrupt rise in interest in the field of the minimal cell. Perhaps the most general reason for this rise of interest lies in a diffuse sense of confidence that the minimal cell is an experimentally accessible target.

Questions for the reader

1. Are you confident that semi-synthetic living cells can be created in the laboratory on the basis of only 30 or 40 genes? If not, can you give some scientific reasoning (as opposed to dubitative feelings) on why not?

2. There are artificial life approaches to the minimal cell, with the idea of creating forms of life other than those based on nucleic acids and proteins. Do you believe that this is possible? Towards which structures would you move?

3. Do you think that the construction of minimal cells may give rise to possible hazards and bioethical problems?

Outlook

The field of the origin of life has progressed very much from the time of Stanley Miller's first experiment. However, the main hypothesis, that cellular life derives from inanimate matter, has not been demonstrated yet. It must then be considered still a working hypothesis. Not that we have alternatives within the realms of science, and I have outlined in Chapter 1 why divine creation cannot be considered as an alternative within science. Of course the question of God is not one that is solved in terms of rationality, but in terms of faith, and we are back to zero.

I have stressed in this book that one of the main reasons why the bottom-up approach to the transition to life has not been corroborated experimentally lies, among others, in the fact that the sequence of our macromolecules of life – enzymes, RNA, and DNA – are the products of the vagaries of contingency and by definition it is then impossible to reproduce them in the laboratory.

I would like to add that there are so many claims about the origin of life – it has been "found" on hydrothermal vents, on ice, on clay, on pyrite, at very high and very low pressure . . . and there are so many corresponding "worlds" – but, as we have seen, all these worlds stop at the synthesis of low-molecular-weight compounds or at the most short oligomers. Of course, for the "origin of life" you need to start from low-molecular-weight compounds, and in this sense the synthesis of water from hydrogen and oxygen can also be considered the origin of life. However, as is mentioned throughout the book, you could have all the low-molecular-weight compounds in any quantity – and you would not be able to make life.

I believe that the bottom-up approach to the origin of life will enjoy a considerable boost, when conditions are found for the prebiotic synthesis of many identical copies of long (>30) co-oligopeptide or co-oligonucleotide sequences. This would at least show that the prebiotic synthesis of enzymes and/or RNA is in principle possible. This remains the main problem with the "prebiotic" RNA world.

Until now, most of the RNA-world literature has been assuming – tacitly or unconsciously – that ribozymes or self-replicating RNA were there to start with – thus constructing the building from the roof down. This tendency is changing, except for a handful of theoreticians who still love to have RNA and ribozymes popping out from nothingness. There is now the search for a pre-RNA world, and the positive tendency to integrate this with the compartmentalist view. This combination is indeed a very helpful development for the field.

More generally, it seems to me that the new wind of system biology may bring about a shift in the question "what is life?;" from a mechanistic viewpoint (nucleic acid mechanisms) to a more integrative view, where self-organization, emergence, and integrative processes may play a major role. I am confident also that this will bring a reappraisal of the concept of autopoiesis and a more philosophical perspective of life science. This development should go hand in hand with a greater sensibility of the newer generation of scientists for the philosophical framework – to which I dedicated some space in this book – including the "bridge" between science and humanistic issues, such as cognition, perception, consciousness, and ethics.

The new Zeitgeist that I mentioned in the preface, connected as it is to system biology, might play an important role in this novel way of looking at this field of science. System biology has many connotations, of which the operative arm is the new field of synthetic biology, a term that I have used in my title. With regard to this I have referred in this book only to a few aspects of synthetic biology, illustrated in Chapter 5, on self-organization, and then more specifically in the description of the project of the minimal cell and that of the "never-born proteins".

This is a highly reduced view of the new discipline of synthetic biology, a term that nowadays encompasses a large variety of approaches to make models of living systems, or part of them, either in terms of molecular biology, or in terms of artificial life – see for example Sismour and Benner, 2005; or the already cited example of Paul and Joyce, 2004. Also biotechnological tools such as chips and DNA arrays are considered part of this new field, which actually, more than a new field, is a re-shaping of somewhat older concepts; but tackled with the ingenuity of new technology. To get a view of the spectrum of the field of synthetic biology, see for example some web addresses.[1]

Going back to the synthetic biology discussed in this book, I mentioned also that the construction in the laboratory of the early cell using the bottom-up approach is made difficult by the clouds of contingency, and added that there is instead confidence in a different approach to the minimal cell. This is the semi-synthetic approach seen in the previous chapter, which utilizes extant enzymes and/or genes.

[1] www.lbl.gov/pbd/synthbio/; www.lbl.gov/pbd/synthbio/default.htm; www.blog.lib.umn.edu/sali0090/synbio; www.stanford.edu/group/kool/synthbio; www.research.dfci.harvard.edu/silverlab/research/synthbio.htm; www.web.mit.edu/synbio/release/conference/synthbio.html; www.austin.che.name/synbio/; www.nature.com/news/2004/041004pf/431624a_pf.html; www.sciencemag.org/cgi/content/full/303/5655/159; www.icos.ethz.ch/news/igem-2005-21k.

It is difficult to make predictions and therefore to share completely the optimism of some of these researchers, but certainly this is a fascinating avenue of work; and one avenue that also has important conceptual implications. It may bring an answer to the questions of whether life is possible with less complexity, whether life is indeed an emergent property arising from the non-living, whether in the history of cell evolution final living cells had indeed as precursors "limping" half-living cells . . .

Even if we are not able to explain how life originated on Earth, we may be able to give a good answer to such questions. This is satisfactory enough and a great motivation for the next generation of life scientists.

References

Abel, D. L. (2002). Is life reducible to complexity? In *Fundamentals of Life*, eds. G. Palyi, C. Zucchi, and L. Caglioti. Elsevier, pp. 57–72.

Achilles, T. and von Kiedrowski, G. (1993). A self-replicating system from three starting materials *Angew. Chem.*, **32**, 1198–201.

Alberts, B., Bray, D., Lewis, J., *et al.* (1989). *Molecular Biology of the Cell*, 2nd edn. Garland Publications Incorporate.

Alexander, S. (1920). *Space, Time and Deity*. McMillan.

Allison, A. C. and Gregoriadis, G. (1974). Liposomes as immunological adjuvants. *Nature*, **252**, 252–8.

Ambartsumian, T. G., Adamian, S. Y., Petrosia, L. S., and Simonian, A. L. (1992). Incorporation of water-soluble enzymes glucose-oxidase and urate oxidase into phosphatidylcholine liposomes. *Biol. Membr.*, **5**, 1878–87.

Anderson, G. and Luisi, P. L. (1979). Papain-induced oligomerization of alpha amino acid esters. *Helv. Chim. Acta*, **62**, 488– 94.

Angelova, M. I. and Dimitrov, D. S. (1988). A mechanism of liposome electro-formation. *Progr. Colloid Polymer. Sci.*, **76**, 59–67.

Annesini, M. C., Braguglia, C. M., Memoli, A., Palermiti, L. G., and Di Sario, S. (1997). Surfactant as modulating agent of enzyme-loaded liposome activity. *Biotechnol. Bioeng.*, **55**, 261–6.

Annesini, M. C., Di Giulio, A., Di Marzio, L., Finazzi-Agrò, A., and Mossa, G. (1992). *J. Liposome Res.*, **2**, 455–67.

Annesini, M. C., Di Giorgio, L., Di Marzio, L., *et al.* (1993). *J. Liposome Res.*, **3**, 639–48.

Annesini, M. C., Di Marzio, L., Finazzi-Agrò, A., Serafino, A. L., and Mossa, G. (1994). Interaction of cationic phospholipid-vesicles with carbonic anhydrase. *Biochem. Mol. Biol. Int.*, **32**, 87–94.

Apte, P. (2002). Vedantic view of Life. In *Fundamentals of Life*, eds. G. Palyi, C. Zucchi, and L. Caglioti. Elsevier, pp. 497–502.

Arinin, E. I (2002). Essence of organic life in Russian orthodox and modern philosophical tradition: beyond functionalism and elementarism. In *Fundamentals of Life*, eds. G. Palyi, C. Zucchi, and L. Caglioti. Elsevier, pp. 503–16.

Ashkenasy, G., Jagasia, R., Yadav, M., and Ghadiri, M. R. (2004). Design of a directed molecular network. *Proc. Natl. Acad. Sci.*, **101**, 10872–7.

Atkins, P. and de Paula, J. (2002). *Physical Chemistry*, 7th edn. Oxford University Press.

Atmanspacher, H. and Bishop, R. (2002). *Between Chance and Choice, Interdisciplinary Perspectives on Determinism*. Imprint Academic.

Avetisov, V. V. and Goldanskii, V. I. (1991). Homochirality and stereospecific activity: evolutionary aspects. *Biosystems*, **25** (3), 141–9.

Ayala, F. J. (1983). Beyond Darwinism? The challenge of macroevolution to the synthetic theory of evolution. In *PSA 1982: Proceedings of the 1982 Biennial Meeting of the Philosophy of Science Association Symposia*, ed. P. D. Asquith and T. Nickles, vol. 2, pp. 275–92.

Baas, N. A. (1994). Emergence, hierarchies, and hyperstructures. In *Artificial Life III, Santa Fe Studies in the Science of Complexity*, ed. C. G. Langton, vol. XVII Addison-Wesley, pp. 515–537.

Bachmann, P. A., Luisi, P. L., and Lang, J. (1992). Autocatalytic self-replication of micelles as models for prebiotic structures. *Nature*, **357**, 57–9.

Bachmann, P. A., Walde, P., Luisi, P. L., and Lang, J. (1990). Self-replicating reverse micelles and chemical autopoiesis. *J. Am. Chem. Soc.*, **112**, 8200–1.

(1991). Self-replicating micelles: aqueous micelles and enzymatically driven reactions in reverse micelles. *J. Am. Chem. Soc.*, **113**, 8204–9.

Bada, J. L. (1997). Meteoritics – extraterrestrial handedness? *Science*, **275**, 942–3.

Bada, J. F. and Lazcano, A. (2002). Some like it hot, but not the first biomolecules. *Science*, **296**, 1982–3.

Bada, J. L. and Lazcano, A. (2003). Prebiotic soup – revisiting the Miller experiment. *Science*, **300**, 745–6.

Baeza, I., Ibáñez, M., Santiago, J. C., *et al.* (1990). Diffusion of Mn^{2+} ions into liposomes mediated by phosphatidate and monitored by the activation of an encapsulated enzymatic system. *J. Mol. Evol.*, **31**, 453–61.

Baeza, I., Wong, C., Mondragón, R., *et al.* (1994). Transbilayer diffusion of divalent cations into liposomes mediated by lipidic particles of phosphatidate, *J. Mol. Evol.*, **39**, 560–8.

Bain, A. (1870). *Logic,* Books II and III. Longmans, Green & Co.

Bak, P., Tang, C., and Wisenfeld, K. (1988). Self-organized criticality. *Physical Rev. A*, **38**, 364–74.

Barbaric, S. and Luisi, P. L. (1981). Micellar solubilization of biopolymers in organic solvents. 5. Activity and conformation of α-chymorypsin in isooctane-AOT reverse micelles. *J. Am. Chem. Soc.*, **103**, 4239–44.

Barrow, J. D. (2001). Cosmology, life and the anthropic principle. *Ann. NY Acad. Sci.*, **950**, 139–53.

Barrow, J. D. and Tipler, F. J. (1986). *The Anthropic Cosmological Principle*. Oxford University Press.

(1988). Action principles in nature. *Nature*, **331**, 31–4.

Bedau, M. A. (1997). Weak emergence. In *Philosophical Perspectives: Mind, Causation and World*, ed. J. Tomberlin. Malden: Blackwell, vol. 11, pp. 375–99.

Ben Jacob, E., Becker, I., Shapira, Y., and Levine, H. (2004). Bacterial linguistic communication and social intelligence. *Trends Microbiol.*, **12**, 366–72.

Benner, S. A. and Sismour, A. M. (2005). Synthetic biology. *Nature Rev. Gen.*, **6**, 524–45.

Berclaz, N., Blöchliger, E., Müller, M., and Luisi, P. L. (2001a). Matrix effect of vesicle formation as investigated by cryotransmission electron microscopy. *J. Phys. Chem. B*, **105**, 1065–71.

Berclaz, N., Müller, M., Walde, P., and Luisi, P. L. (2001b). Growth and transformation of vesicles studied by ferritin labeling and cryotransmission electron microscopy. *J. Phys. Chem. B*, **105**, 1056–64.

Bernal, J. D. (1951). *The Physical Basis of Life*. Routledge & Paul.

(1965), in *Theoretical and Mathematical Biology*, eds. T. H. Waterman and H. J. Morowitz. Blaisdell.

(1967). *The Origin of Life*. World Publishing Company.

(1971). *Der Ursprung des Lebens*. Editions Rencontre.

Bernard, C. (1865). *Introduction to the Study of Experimental Medicine*. Translated by H. C. Greene (1927). Henry Schuman.

Berti, D., Luisi, P. L., and Baglioni, P. (2000). Molecular recognition in supramolecular structures formed by phosphatidylnucleosides-based amphiphiles. *Colloids Surf. A*, **167**, 95–103.

Berti, D., Baglioni, P., Bonaccio, S., Barsacchi-Bo, G., and Luisi, P. L. (1998). Base complementarity and nucleoside recognition in phosphatidylnucleoside vesicles. *J. Phys. Chem. B*, **102**, 303–8.

Bianucci, M., Maestro, M., and Walde, P. (1990). Bell-shaped curves of the enzyme-activity in reverse micelles – a simplified model for hydrolytic reactions. *Chem. Phys.*, **141**, 273–83.

Biebricher, K., Eigen, M., and Luce, R. (1981). Kinetic analysis of template, instructed and *de novo* RNA synthesis by Qbeta replicase. *J. Mol. Biol.*, **148**, 391–410.

Billmeyer, F. W. (1984). *Textbook of Polymer Science*, 3rd edn. Wiley & Sons.

Birdi, K. S. (1999). *Self-Assembly Monolayer Structures of Lipids and Macromolecules At Interfaces*. Plenum Press.

Bissel, R. A., Cordova, E., Kaifer, A. E., and Stoddart, J. F. (1994). A chemically and electrochemically switchable molecular shuttle. *Nature*, **369**, 133.

Bitbol, M. (2001). Non-representationalist theories of knowledge and quantum mechanics. *SATS, Nordic J. Phil.*, **2**, 37–62.

Bitbol, M. and Luisi, P. L. (2004). Autopoiesis with or without cognition: defining life at its edge. *J. Royal. Soc. Interface*, **1**, 99–107.

Blocher, M., Hitz, T., and Luisi, P. L. (2001). Stereoselectivity in the oligomerization of racemic Tryptophan *N*-Carboxyanhydride (NCA-Trp) as determined by isotopic labelling and mass spectrometry. *Helv. Chim. Acta*, **84**, 842–8.

Blocher, M., Liu, D., and Luisi, P. L. (2000). Liposome-assisted selective polycondensation of α-amino acids and peptides: the case of charged liposomes. *Macromolecules*, **33**, 5787–96.

Blocher, M., Walde, P., and Dunn, I. J. (1999). Modeling of enzymatic reactions in vesicles: the case of alpha-chymotrypsin. *J. Biotechnol. Bioeng.*, **62**, 36–43.

Blöchiger, E., Blocher, M., Walde, P., and Luisi, P. L. (1998). Matrix effect in the size distribution of fatty acid vesicles. *J. Phys. Chem.*, **102**, 10383–90.

Böhler, C., Bannwarth, W., and Luisi, P. L. (1993). Self-replication of oligonucleotides in reverse micelles. *Helv. Chim. Acta*, **76**, 2313–20.

Böhringer, M., Morgenstern, K., Schneider, W. D., and Berndt, R. (1999). Separation of a racemic mixture of two-dimensional molecular clusters by scanning tunneling microscopy. *Angew. Chem., Int. Ed. Engl.*, **38**, 821–3.

Boiteau, L., Plasson, R., Collet, H., *et al.* (2002). Molecular origin of life: when chemistry became cyclic. The primary pump, a model for prebiotic emergence and evolution of petides. In *Fundamentals of Life*, eds. G. Palyi, C. Zucchi, and L. Caglioti. Elsevier, pp. 211–18.

Boicelli, C. A., Conti, F., Giomini, M., and Giuliani, A. M. (1982). Interactions of small molecules with phospholipids in inverted micelles. *Chem. Phys. Lett.*, **89**, 490–6.

Bolli, M., Micura, R., and Eschenmoser, A. (1997a). Pyranosyl-RNA: chiroselective self-assembly of base sequences by ligative oligomerization of tetranucleotide-2′,3′-cyclophosphates (with a commentary concerning the origin of biomolecular homochirality). *Chem. Biol.*, **4**, 309–20.

Bolli, M., Micura, R., Pitsch, S., and Eschenmoser, A. (1997b). Pyranosyl-RNA: further observations on replication. *Helv. Chim. Acta*, **80**, 1901–51.

Bonaccio, S., Walde, P., and Luisi, P. L. (1994a). Liposomes containing purine and pyrimidine bases: stable unilamellar liposomes from phosphatidyl nucleosides. *J. Phys. Chem.*, **98**, 6661–3.

Bonaccio, S., Cescato, C., Walde, P., and Luisi, P. L. (1994b). Self-production of supramolecular structures. In *Liposomes from Lipidonucleotides and from Lipidopeptides*, eds G. R. Fleischaker *et al.* Kluwer Academic, pp. 225–59.

Bonaccio, S., Capitani, D., Segre, A. L., Walde, P., and Luisi, P. L. (1997). Liposomes from phosphatidyl nucleosides: an NMR investigation. *Langmuir*, **13**, 1952–6.

Bonaccio, S., Wessicken, M., Berti, D., Walde, P., and Luisi, P. L. (1996). Relation between the molecular structure of phosphatidyl nucleosides and the morphology of their supramolecular and mesoscopic aggregates. *Langmuir*, **12**, 4976–78.

Bonner, W. A. (1999). Chirality amplification – the accumulation principle revisited. *Orig. Life Evol. Biosph.*, **29**, 615–23.

Böttcher, B., Lucken, U., and Graber, P. (1995). The structure of the H^+-ATPase from chloroplasts by electron cryomicroscopy. *Biochem. Soc. Trans.*, **23**, 780–5.

Bourgine P. and Stewart, J. (2004). Autopoiesis and cognition. *Artificial Life*, **10** (3), 327–45.

Brack, A. (ed.) (1998). *The Molecular Origin of Life*. Cambridge University Press.

Brasier, M. D., Green, O. R., Jephcoat, A. P., *et al.* (2002). Questioning the evidence for Earth's oldest fossils. *Nature*, **416**, 76–7.

Briggs, T. and Rauscher, W. (1973). An oscillating iodine clock. *J. Chem. Educ.*, **50**, 496.

Britt, R. R. (2000). Are we all aliens? The new case for panspermia. http://www.space.com.

Broad, C. D. (1925). *The Mind and its Place in Nature*. Routledge and Kegan.

Bucknall, D. G. and Anderson, H. L. (2003). Polymers get organized. *Science*, **302**, 1904–5.

Buhse, T., Lavabre, D., Nagarajan, R., and Micheau, J. C. (1998). Origin of autocatalysis in the biphasic alkaline hydrolysis of C-4 to C-8 ethyl alkanoates. *J. Phys. Chem. A*, **102**, 10552–9.

Buhse, T., Nagarajan, R., Lavabre, D., and Micheau, J. C. (1997). Phase-transfer model for the dynamics of "micellar autocatalysis". *J. Phys. Chem. A*, **101**, 3910–17.

Bujdak, J., Eder, A., Yongyai, Y., Faybikova, K., and Rode, B. M. (1995). Peptide chain elongation: a possible role of montmorillonite in prebiotic synthesis of protein precursors. *Orig. Life Evol. Biosph.*, **5**, 431–41.

Bujdak, J., Slosiarikova, H., Texler, N., Schwendinger, M., and Rode, B. M. (1994). On the possible role of montmorillonites in prebiotic peptide formation. *Monats. Chem.*, **125**, 1033–9.

Burmeister, J. (1998). Self-replication and autocatalysis. In *The Molecular Origin of Life*, ed. A. Brack. Cambridge University Press, pp. 295–310.

Cairns-Smith, A. G. (1977). Takeover mechanisms and early biochemical evolution. *Biosystems*, **9**, 105–9.

(1978). Precambrian solution photochemistry, inverse segregation, and banded iron formations. *Nature*, **276**, 808–9.

(1982). *Genetic Takeover and the Mineral Origins of Life*. Cambridge University Press.

(1990). *Seven Clues to the Origin of Life*. Cambridge University Press.

Cairns-Smith, A. G. and Walker, G. L. (1974). Primitive metabolism. *Curr. Mod. Biol.*, **5** (4), 173–86.

Cairns-Smith, A. G., Hall, A. J., and Russell, M. J. (1992). Mineral theories of the origin of life and an iron sulphide example. *Orig. Life Evol. Biosph.*, **22**, 161–80.

Calderone, C. T. and Liu, D. R. (2004). Nucleic acid-templated synthesis as a model system for ancient translation. *Curr. Opin. Chem. Biol.*, **8**, 645–53.

Capra, F. (2002). *The Hidden Connections*. Harper Collins.

Carey, M. V. and Small, D. M. (1972). Micelle formation by bile salts. Physical-chemical and thermodynamic considerations. *Arch. Intern. Med.*, **130**, 506–27.

Carr, B. (2001). Life, the cosmos and everything. *Phys. World*, **14**, 23–5.

Caselli, M., Maestro, M., and Morea, G. (1988). A simplified model for protein inclusion in reverse micelles. SANS measurements as a control test. *Biotech. Prog.*, **4**, 102–6.

Cello, J., Paul, A. V., and Wimmer, E. (2002). Chemical synthesis of poliovirus cDNA: generation of infectious virus in the absence of natural template. *Science*, **297**, 1016–18.

Celovsky, V. and Bordusa, F. (2000). Protease-catalyzed fragment condensation via substrate mimetic strategy: a useful combination of solid-phase peptide synthesis with enzymatic methods. *J. Pept. Res.*, **55**, 325–9.

Cevc, G. (1992). In *Liposome Dermatics*, eds. O. Braun-Falco, H. C. Korting, and H. I. Maibach. Springer Verlag, pp. 82–90.

Chakrabarti, A. C., Breaker, R. R., Joye, G. F., and Deamer, D. W. (1994). Production of RNA by a polymerase protein encapsulated within phospholipid vesicles. *J. Mol. Evol.*, **39**, 555–9.

Chapman, K. B. and Szostak, J. W. (1995). Isolation of a ribozyme with 5'-5' ligase activity. *Chem. Biol.*, **2**, 325–33.

Chen, I. A. and Szostack, J. W. (2004). A kinetic study of the growth of fatty acid vesicles. *Bioph. J.*, **87**, 988–98.

Chen, I. A., Roberts, R. W., and Szostak, J. W. (2004). The Emergence of competition between model protocells. *Science*, **305**, 1474–6.

Cheng, Z. and Luisi, P. L. (2003). Coexistence and mutual competition of vesicles with different size distributions. *J. Phys. Chem. B*, **107** (39), 10940–5.

Christidis, T. (2002). Probabilistic causality and irreversibility: Heraclitus and Prigogine. In *Between Chance and Choice*, eds. H. Atmanspacher and R. Bishop. Academic Imprint.

Chyba, C. F. and Sagan, C. (1992). Endogenous production, exogenous delivery and impact-shock synthesis of organic molelcules: an inventory for the origin of life. *Nature*, **355**, 125–32.

Chyba, F. and McDonald, G. D. (1995). The origin of life in the solar system: current issues. *Ann. Rev. Earth Planet. Sci.*, **23**, 215–49.

Cistola, D. P., Hamilton, J. A., Jackson, D., and Small, D. M. (1988). Ionization and phase-behavior of fatty-acids in water. Application of the Gibbs phase rule. *Biochemistry*, **27**, 1881–8.

Commeyras, A., Boiteau, L., Vandenabeele-Trambouze, O., and Selsis, F. (2005). From prebiotic chemistry to the origins of life on Earth. In *Lectures in Astrobiology*, eds. M. Gargaud, B. Barbier, H. Martin and J. Reisse. Springer-Verlag, vol. I, part II, pp. 35–55.

Conway-Morris, S. (2003). *Life's Solution, Inevitable Humans in a Lonely Universe*. Cambridge University Press.

Cooper, G. W., Onwo, W. M., and Cronin, J. R. (1992). Alkyl phosphonic acids and sulfonic acids in the Murchison meteorite. *Geochim. cosmochim. acta*, **56**, 4109–15.

Corliss, J. B., Baross, J. A., and Hoffman, S. E. (1981). An hypothesis concerning the relationship between submarine hot springs and the origin of life. *Oceanologica acta 4 Suppl.*, 59–69.

Corrin, M. L., Klevens, H. B., and Harkins, D. (1946). The determination of critical concentrations for the formation of soap micelles by the spectral behavior of pinacyanol chloride. *J. Chem. Phys.*, **14**, 480–6.

Coveney, P. and Highfield, R. (1990). *The Arrow of Time*. W. H. Allen.

Crick, F. (1966). *Of Molecules and Men*. University of Washington Press.
  (1980). *The Astonishing Hypothesis. The Search of the Soul from a Chemical Perspective*. Scribner.

Cronin, J. R. and Pizzarello, S. (1997). Enantiomeric excesses in meteoritic amino acids. *Science*, **275**, 951–5.

Crusats, J., Claret, J., Díez-Pérez, I., *et al.* (2003). Chiral shape and enantioselective growth of colloidal particles of self-assembled meso-tetra(phenyl and 4-sulfonatophenyl)porphyrins. *Chem. Commun.*, **13**, 1588–9.

Cullis, P. R., Hope, M. J., Bally, M. B., *et al.* (1987). Liposomes as pharmaceuticals. In *Liposomes. From Biophysics to Therapeutics*, ed. M. J. Ostro. Marcel Dekker, pp. 39–72.

Cullis, P. R., Hope, M. J., and Tilcock, C. P. S. (1986). Lipid polymorphism and the role of lipids in membranes. *Chem. Phys. Lipids*, **40**, 127–44.

Damasio, A. R. (1999). *The Feeling of What Happens*. Harcourt.

Davies, P. (1999). *The Fifth Miracle: The Search for the Origin and Meaning of Life*. Simon & Schuster.

Dawkins, R. (1990). *The Blind Watchmaker: Why the Evidence of Evolution Reveals a Universe without Design*. Penguin Books.
  (2002). *How Life Began: The Genesis of Life on Earth*. Foundation for New Directions.

Deamer, D. W. (1985). Boundary structures are formed by organic components of the Murchison carbonaceous chondrite. *Nature*, **317**, 792–4.
  (1998). Possible starts for primitive life. In *The Molecular Origins of Life*, ed. A. Brack. Cambridge University Press.

Deamer, D. W. and Pashley, R. M. (1989). Amphiphilic components of the Murchison carbonaceous chondrite: surface properties and membrane formation. *Orig. Life Evol. Biosph.*, **19**, 21–38.

Deamer, D. W., Harang-Mahon, E., and Bosco, G. (1994). Self-assembly and function of primitive membrane structures. In *Early Life on Earth. Nobel Symposium No. 84*, ed. S. Bengston. Columbia University Press, pp. 107–123.

Decher, G. (1997). Fuzzy nano-assemblies: toward layered polymeric multicomposites. *Science*, **277**, 1232–7.

Decker, P., Schweer, H., and Pohlmann, R. (1982). Identification of formose sugars, presumable prebiotic metabolites, using capillary gas chromatography/gas chromatography-mass spectrometry of n-butoxime trifluoroacetates on OV-225. *J. Chromatogr.*, **225**, 281–91.

de Duve, C. (1991). *Blueprint for a Cell : The Nature and the Origin of Life*. Neil Patterson Publishers.
  (2002). *Life Evolving: Molecules, Mind and Meaning*. Oxford University Press.

de Duve, C. and Miller, S. (1991). Two-dimensional life? *Proc. Natl. Acad. Sci.*, **88**, 10014–17.

de Feyter, S., Gesquiere, A., Wurst, K., *et al.* (2001). Homo- and heterochiral supramolecular tapes from achiral, enantiopure, and racemic promesogenic formamides: expression of molecular chirality in two and three dimensions. *Angew. Chem. Int. Ed. Eng.*, **40**, 3217–20.

de Kruijff, B., Cullis, P. R. and Verkleij, A. J. (1980). Non-bilayer lipid structures in model and biological membranes. *Trends Bioch. Sci.*, **5**, 79–81.

de Napoli, M., Nardis, S., Paolesse, R., (2004). Hierarchical porphyrin self-assembly in aqueous solution. *J. Am. Chem. Soc.*, **126**, 5934–5.

Ding, P. Z., Kawamura, K., and Ferris, J. P. (1996). Oligomerization of uridine phosphorimidazolides on montmorillonite: a model for the prebiotic synthesis of RNA on minerals. *Orig. Life Evol. Biosph.*, **26**, 151–71.

Domazou, A. S. and Luisi, P. L. (2002). Size distribution of spontaneously formed liposomes by the alcohol injection method. *J. Liposome Res.*, **12** (3), 205–20.

Douglas, S., Zauner, S., Fraunholz, M., *et al.* (2001). The highly reduced genome of an enslaved algal nucleus. *Nature*, **410**, 1091–2.

Dubois, L. H. and Nuzzo, R. G. (1992). Synthesis, structure, and properties of model organic-surfaces. *Ann. Rev. Phys. Chem.*, **43**, 437–63.

Dworkin, J. D., Deamer, D. W., Sandford, S., and Allmandola, L. (2001). Self-assembling amphiphilic molecules: synthesis in simulated interstellar/precometary ices. *Proc. Natl. Acad. Sci.*, **98**, 815–19.

Dyson, F. J. (1982). A model for the origin of life. *J. Mol. Evol.*, **18**, 344–50.
  (1985). *Origins of Life*. Cambridge University Press.

Ehrenfreund, P., Irvine, W., Becker, L., *et al.* (2002). Astrophysical and astrochemical insights into the origin of life. *Rep. Prog. Phys.*, **65**, 1427–87.

Eichhorn, U., Bommarius, A. S., Drauz, K., and Jakubke, H.-D. (1997). Synthesis of dipeptides by suspension-to-suspension conversion via thermolysin catalysis: from analytical to preparative scale. *J. Pept. Sci.*, **3**, 245–51.

Eigen, M. (1971). Self-organization of matter and the evolution of biological macromolecules. *Naturwissenschaften*, **58**, 465–523.

Eigen, M. and Schuster, P. (1977). Hypercycle – principle of natural self-organization. A. Emergence of hypercycle. *Naturwissenschaften*, **64**, 541–65.
  (1979). *The Hypercycle: a Principle of Natural Self-Organization*. Springer Verlag.

Eigen, M. and Winkler-Oswatitisch, R. (1992). *Steps Towards Life*. Oxford University Press.

El Seoud, O. A. (1984). In *Reverse Micelles*, eds. P. L. Luisi and B. Straub. Plenum Press.

Elias, H. G. (1997). *An Introduction to Polymer Science*. Wiley VCH.

Engels, F. (1894). In *Herrn Eugen Dühring's Umwalzung der Wissenschaft*, Dietz Verlag.

Epstein, S., Krishnamurthy, R. V., Cronin, J. R., Pizzarello, S., and Yuen, G. U. (1987). Unusual stable isotope ratios in amino acid and carboxylic-acid extracts from the Murchison meteorite. *Nature*, **326**, 477–9.

Erickson, J. C. and Kennedy R. M. (1980). Effects of histidyl-histidine and polyribonucleotides on glycine condensation in fluctuating clay environments. *Abstracts Papers Am. Chem. Soc.*, **179**, 43.

Ericsson, B., Larsson, K., and Fontell, K. (1983). A cubic protein-monoolein-water phase. *Biochim. Biophys. Acta*, **729**, 23–7.

Erwin, D. H. (2003). Life's solution – inevitable humans in a lonely universe. *Science*, **302**, 1682–3.

Eschenmoser, A. (1999). Chemical etiology of nucleic acid structure. *Science*, **284**, 2118–24.
  (2003). Creating a perspective for comparing. In *Proceedings of the J. Templeton Foundation "Biochemistry and Fine-tuning"*, Harvard University, October 10–12, 2003.

Eschenmoser, A. and Kisakürek, M. V. (1996). Chemistry and the origin of life. *Helv. Chim. Acta*, **79**, 1249–59.

Fadnavis, N. W. and Luisi, P. L. (1989). Immobilized enzymes in reverse micelles: studies with gel-entrapped Trypsin and alpha-Chymotrypsin in AOT reverse micelles. *Biotechnol. Bioeng.*, **33**, 1277–82.

Falbe, J. (1987). *Surfactants in Consumer Products. Theory, Technology and Applications.* Springer Verlag.

Famiglietti, M., Hochköppler, A., and Luisi, P. L. (1993). Surfactant-induced hydrogen production in cyanobacteria. *Biotechnol. Bioeng.*, **42**, 1014–18.

Famiglietti, M., Hochköppler, A., Wehrli, E., and Luisi, P. L. (1992). Photosynthetic activity of cyanobacteria in water-in-oil microemulsions. *Biotechnol. Bioeng.*, **40**, 173–8.

Farre, L. and Oksala, T., eds. (1998). Emergency, complexity, hierarchy, organisation. Selected papers from the ECHO III Conference (ESPOO, Finland), *Acta Polytechnica Scandi.*, **91**.

Fendler J. H. and Fendler, E. J. (1975). *Catalysis in Micellar and Macromolecular Systems.* Academic Press.

Ferris, J. P. (1998). Catalyzed RNA synthesis for the RNA world. In *The Molecular Origin of Life*, ed. A. Brack. Cambridge University Press, pp. 255–6.

(2002). From building blocks to the polymers of life. In *Life's Origin, The Beginning of Biological Evolution*, ed. J. W. Schopf. California University Press, 113–39.

Ferris, J. P. and Ertem, G. (1992). Oligomerization reaction of ribonucleosides on montmorillonite: reaction of 5′-phosphorimidazolide of adenosine. *Science*, **257**, 1387–9.

(1993). Montmorillonite catalysis of RNA oligomer formation in aqueous solution: a model for the prebiotic formation of RNA. *J. Am. Chem. Soc.*, **115**, 12270–5.

Ferris, J. P., Donner, D. B., and Lobo, A. P. (1973). Possible role of hydrogen cyanide in chemical evolution. The oligomerization and condensation of hydrogen cyanide. *J. Mol. Biol.*, **74**, 511–18.

Ferris, J. P., Sanchez, R. A., and Orgel, L. E. (1968). Studies in prebiotic synthesis. III, Synthesis of pyrimidines from cyanoacetilene and cyanate. *J. Mol. Biol.*, **33**, 693–704.

Ferris, J. P., Hill, R. Jr., Liu, R., and Orgel, L. (1996). Synthesis of long prebiotic oligomers on mineral surface. *Nature*, **381**, 59–61.

Ferris, J. P., Joshi, P. C., Edelson, E. H., and Lawless, J. G. (1978). HCN: a plausible source of purines, pyrimidines and amino acids on the primitive earth. *J. Mol. Evol.*, **11**, 293–311.

Ferris, J. P., Wos, J. D., Nooner, D. W., and Oró, J. (1974). Chemical Evolution. 21. Amino-Acids Released on Hydrolysis of HCN Oligomers. *J. Mol. Evol.*, **3**, 225–31.

Field, R. J. (1972). A reaction periodic in time and space. *J. Chem. Educ.*, **49**, 308–11.

Fischer, A., Franco, A., and Oberholzer, T. (2002). Giant vesicles as microreactors for enzymatic mRNA synthesis. *Chem. Bio. Chem.*, **3** (5), 409–17.

Fischer, A., Oberholzer, T., and Luisi, P. L. (2000). Giant vesicles as models to study the interactions between membranes and proteins. *Biochim. Biophys. Acta*, **1467**, 177–88.

Fleischaker, G. (1988). Autopoiesis: the status of its system logic. *Biosystems*, **22**, 37–49.

Fletcher, P. D. and Robinson, B. H. (1981). *Ber. Bunsenges. Phys. Chem.*, **85**, 863.

Foldvari, M., Geszles, A., and Mezei, M. (1990). *J. Microencapsul.*, **7**, 479–89.

Folsome, C. E. (1979). *The Origin of Life: A Warm Little Pond.* W. H. Freeman & Co.

Fontell, K. (1990). Cubic phases in surfactant and surfactant-like lipid systems. *Colloid Polym. Sci.*, **268**, 265–85.

Föster, S. and Plantenberg, T. (2002). From self-organizing polymers to nanohybrid and biomaterials. *Angew. Chem. Int. Ed. Engl.*, **41**, 688–714.

Fox, S. W. (1988). *The Emergence of Life*. Basic Books.

Fox, S. W. and Dose, K. (1972). *Molecular Evolution and the Origin of Life*. W. H. Freeman.

(1977). *Molecular Evolution and the Origin of Life*. New York and Basel: Marcel Dekker.

Franchi, M. and Gallori, E. (2004). Origin, persistence and biological activity of genetic material in prebiotic habitats. *Orig. Life Evol. Biosph.*, **34**, 133–41.

Franchi, M., Ferris, J. P., and Gallori, E. (2003). Cations as mediators of the adsorption of nucleic acids on clay surfaces in prebiotic environments. *Orig. Life Evol. Biosph.*, **33**, 1–16.

Franz, M.-L. von (1988). *Psyche und Materie*. Daimon Verlag.

Fraser, C. M., Gocayne, J. D., White, O., *et al.* (1995). The minimal gene complement of *Mycoplasma genitalium*. *Science*, **270**, 397–403.

Freitas, R. A., Jr. and Merkle, R. C. (2004). *Kinematic Self-Replicating Machines*. Landes Bioscience.

Frick, D. N. and Richardson, C. C. (2001). DNA Primases. *Annu. Rev. Biochem.*, **70**, 39–80.

Funqua, C., Parsek, M. R., and Greenberg, E. P. (2001). Regulation of gene expression by cell-to-cell communication: acyl-homoserine lactone quorum sensing. *Ann. Rev., Genet.*, **35**, 439–68.

Ganti, T. (1975). Organization of chemical reactions into dividing and metabolizing units: the chemotons. *BioSystems*, **7**, 15–21.

(1984). *Chemoton elmélet 1. kötet. A fluid automaták elméleti alapjai* (Translated as Chemoton Theory. Vol. 1. Theory of Fluid Automata). OMIKK.

(2003). *The Principles of Life*. Oxford University Press.

Gao, X. and Huang, L. (1995). Cationic liposome-mediated gene transfer, *Gene Ther.*, **2** (10), 710–22.

Gavrilova, L. P., Kostiashkina, O. E., Koteliansky, V. E., Rutkevitch, N. M., and Spirin, A. S. (1976). Factor-free (non-enzymic) and factor-dependent systems of translation of polyuridylic acid by *E. coli* ribosomes. *J. Mol. Biol.*, **101**, 537–52.

Gennis, R. B. (1989). *Biomembranes, Molecular Structure and Function*. Springer Verlag.

Ghosh, I. and Chmielewski, J. (2004). Peptide self-assembly as a model of proteins in the pre-genomic world. *Curr. Opin. Chem. Biol.*, **8**, 640–4.

Gil, R., Silva, F. J., Peretó, J., and Moya, A. (2004). Determination of the core of a minimal bacteria gene set. *Microb. Molec. Biol. Rev.*, **68**, 518–37.

Gil, R., Sabater-Munoz, B., Latorre, A., Silva, F. J., and Moya, A. (2002). Extreme genome reduction in *Buchnera* spp: toward the minimal genome needed for symbiotic life. *Proc. Natl. Acad. Sci. USA*, **99**, 4454–8.

Gilbert, W. (1986). The RNA world. *Nature*, **319**, 618.

Glade, N., Demongeot, J., and Tabony, J. (2004). Microtubules self-organization by reaction-diffusion processes causes collective transport and organization of cellular particles. *BMC Cell Biol.*, **5**, 5–23.

Glotzer, S. C. (2004). Materials science. Some assembly required. *Science*, **306**, 419–20.

Gould, S. J. (1989). *Wonderful Life*. Penguin Books.

Graf, A., Winterhalter, M., and Meier, W. (2001). Nanoreactors from polymer-stabilized liposomes. *Langmuir*, **17**, 919–23.

Gregoriadis G. (ed.) (1988). *Liposomes ad Carriers of Drugs: Recent Trends and Progress*. Wiley.

(1976a). The carrier potential of liposomes in biology and medicine (first of two parts). *New Engl. J. Med.*, **295**, 704–10.

(1976b). The carrier potential of liposomes in biology and medicine (second of two parts). *New Engl. J. Med.*, **295**, 765–70.

(1995). Engineering liposomes for drug delivery: progress and problems. *Trends Biotechnol.*, **13** (12), 527–37.

Häckel, E. (1866). *Allgemeine Anatomie der Organismen*. Walter de Gruyer.

Häring, G., Luisi, P. L., and Hauser H. (1988). Characterization by electron spin resonance of reversed micelles consisting of the ternary system AOT–isooctan–water. *J. Phys. Chem.*, **92**, 3574–81.

Häring, G., Luisi, P. L., and Meussdoerffer, F. (1985). Solubilization of bacteria cells in organic solvents via reverse micelles. *Biochem. Biophys. Res. Commun.*, **127**, 911–15.

Häring, G., Pessina, A., Meussdoerffer, F., Hochkoppler, A., and Luisi, P. L. (1987). Solubilization of bacterial cells in organic solvents via reverse micelles and microemulsions. *Ann. Biochem. Eng.*, **506**, 337–344.

Haldane, J. B. S. (1929). The origin of life. *Rationalist Annual*, **148**, 3–10.

(1954). The origin of life. *New Biol.*, **16**, 12–27.

(1967). In *The Origin of Life*, ed. J. D. Bernal. World Publishing Co.

Halling, P. J., Eichhorn, U., Kuhl, P., and Jakubke, H.-D. (1995). Thermodynamics of solid-to-solid conversion and application to enzymic peptide synthesis. *Enzyme Microb. Technol.*, **17**, 601–6.

Han, D. and Rhee, J. S. (1986). *Biotechnol. Bioeng.*, **27**, 1250–5.

Hanczyc, M. M., Fujikawa, S. M., and Szostak, J. W. (2003). Experimental models of primitive cellular compartments: encapsulation, growth, and division. *Science*, **302**, 618–22.

Hansler, M. and Jakubke, H.-D. (1996). Nonconventional protease catalysis in frozen aqueous solutions. *J. Pept. Sci.*, **2**, 279–89.

Harada, S. and Schelly, Z. A. (1982). Reversed micelle of dodecylpyridinium iodide in benzene. Pressure-jump relaxation kinetic and equilibrium study of the solubilization of 7,7,8,8-tetracyanoquinodimethane. *J. Phys. Chem.*, **86**, 2098–102.

Hargreaves, W. R and Deamer, D. W. (1978a). Liposomes from ionic, single-chain amphiphiles. *Biochemistry*, **17**, 3759–68.

(1978b). In *Light Transducing membranes: Structure, Function and Evolution*, ed. D. W. Deamer. Academic Press, pp. 23–59.

Hargreaves, W. R., Mulvhill S. J., and Deamer, D. W. (1977). Synthesis of phospholipids and membranes in prebiotic conditions. *Nature*, **266**, 78–80.

Havinga, E. (1954). Spontaneous formation of optically active substances. *Biochem. Biophys. Acta*, **13**, 171–4.

Hawker, C. J. and Frechet, J. M. J. (1990). Preparation of polymers with controlled molecular architecture – a new convergent approach to dendritic macromolecules. *J. Am. Chem. Soc.*, **112**, 7638–47.

Hayatsu, R., Studier, M. H., Moore, L. P., and Anders, E. (1975). Purines and triazines in the Murchison meteorite. *Geochim. Cosmochim. Acta*, **39**, 471–88.

Heinen, W. and Lauwers, A. M. (1997). The iron-sulfur world and the origins of life: abiotic thiol synthesis from metallic iron, H_2S and CO_2; a comparison of the thiol generating $FeS/HCl(H_2S)/CO_2$-system and its $Fe^0/H_2S/CO_2$-counterpart. *Proc. Royal Netherlands Acad. Arts Sci.*, **100**, 11–25.

Hilborn, R. C. (1994). *Chaos and Non Linear Dynamics*. Oxford University Press.

Hilhorst, R., Spruijt, R., Laane, C., and Veeger, C. (1984). Rules for the regulation of enzyme-activity in reversed micelles as illustrated by the conversion of apolar steroids by 20-beta-hydroxysteroid dehydrogenase. *Eur. J. Biochem.*, **144**, 459–66.

Hitz, T. and Luisi, P. L. (2004). Spontaneous onset of homochirality in oligopeptide chains generated in the polymerization of *N*-carboxyanhydride amino acids in water. *Orig. Life Evol. Biosph.*, **34** (1–2), 93–110.

Hochköppler, A. and Luisi, P. L. (1989). Solubilization of soybean mitochondria in AOT/isooctane water-in-oil microemulsions. *Biotechnol. Bioeng.*, **33**, 1477–81.

 (1991). Photosynthetic activity of plant cells solubilized in water-in-oil microemulsions. *Biotechnol. Bioeng.*, **37**, 918–21.

Hochköppler, A., Pfammatter, N., and Luisi, P. L. (1989). Activity of yeast cells solubilized in water-in-oil microemulsions. *Chimia*, **43**, 348–350.

Holden, C. (2005). Vatican astronomer rebuts cardinals' attack on Darwinism. *Science*, **309**, 996–7.

Holland, J. H. (1998). *Emergence: From Chaos to Order*. Oxford University Press.

Holm, N. G. and Andersson, E. M. (1998). Hydrothermal systems. In *The Molecular Origin of Life*, ed. A. Brack. Cambridge University Press.

Holm, N. G., Cairns-Smith, A. G., Daniel, R. M., *et al.* (1992). Marine hydrothermal systems and the origin of life: future research. *Orig. Life Evol. Biosph.*, **22**, 181–242.

Horowitz, N. and Miller, S. (1962). In *Progress in the Chemistry of Natural Products*, ed. L. Zechmeister. Springer Verlag, vol. 20, pp. 423–59.

Horowitz, P. and Sagan, C. (1993). Five years of Project META: an all-sky narrow-band radio search for extraterrestrial signals. *Astrophys. J.*, **415**, 218–33.

Howard, F. B., Frazier, J., Singer, M. F., and Miles, H. T. (1966). Helix formation between polyribonucleotides and purine nucleosides and nucleotides. 2. *J. Mol. Biol.*, **16**, 415.

Hoyle, F. and Wickramasinghe, C. (1999). Astronomical origins of life – steps toward panspermia. *Astrophys. Space Sci.*, **268**, preface VII–VIII.

Hoyle, F., and Wickramasinghe, C. (2000). *Astronomical Origins of Life – Steps Towards Panspermia*. Kluwer Academic.

Huang, S. S. (1959). Occurrence of life in the universe. *Amer. Sci.*, **47**, 397–402.

Huang, W. M. and. Tso, P. O. P. (1966). Physicochemical basis of recognition process in nucleic acid interactions. I. Interactions of polyuridylic acid and nucleosides. *J. Mol. Biol.*, **16**, 523.

Huber, C. and Wächtershäuser, G. (1997). Activated acetic acid by carbon fixation on (Fe, Ni)S under primordial condition. *Science*, **276**, 245–7.

Hutchinson, C. A., Peterson, S. N., Gill, S. R., *et al.* (1999). Global transposon mutagenesis and a minimal Mycoplasma genome. *Science*, **286**, 2165–9.

Imre, V. E. and Luisi, P. L. (1982). Solubilization and condensed packaging of nucleic acids in reversed micelles. *Biochem. Biophys. Res. Commun.*, **107**, 538–45.

Inoue, T. and Orgel, L. (1983). A nonenzymatic RNA polymerase model. *Science*, **219**, 859–62.

Ishikawa, K., Sato, K., Shima, Y., Urabe, I., and Yomo, T. (2004). Expression of cascading genetic network within liposomes. *FEBS Lett.*, **576**, 387–90.

Islas, S., Becerra, A., Luisi, P. L., and Lazcano, A. (2004). Comparative genomics and the gene complement of a minimal cell. *Orig. Life Evol. Biosph.*, **34** (1–2), 243–56.

Israelachvili, J. N, Mitchell, D. J., and Ninham, B. W. (1977). Theory of self-assembly of lipid bilayers and vesicles. *Biochim. Biophys. Acta*, **470**, 185–201.

Israelachvili, J. N. (1992). *Intermolecular and Surface Forces*, 2nd edn. Academic Press.

Issac, R. and Chmielewski, J. (2002). Approaching exponential growth with a self-replicating peptide. *J. Am. Chem. Soc.*, **124**, 6808–9.

Itaya, M. (1995). An estimation of the minimal genome size required for life. *FEBS Lett.*, **362**, 257–60.

Ito, Y., Fujii, H., and Imanishi, Y. (1993). Catalytic peptide synthesis by trypsin modified with polystyrene in chloroform. *Biotechnol. Prog.*, **9** (2), 128–30.

Itojima, Y., Ogawa, Y., Tsuno, K., Handa, N., and Yanagawa, H. (1992). Spontaneous formation of helical structures from phospholipid-nucleoside conjugates. *Biochemistry*, **31**, 4757–65.

Jacob, F. (1982). *The Possible and the Actual*. University of Washington Press.

Jager, L., Wright, M. C., and Joyce, G. F. (1999). A complex ligase ribozyme evolved in vitro from a group I ribozymes domain, *Proc. Natl. Acad. Sci.*, **96**, 14712–17.

Jakubke, H.-D. (1987). In *The Peptides: Analysis, Synthesis, Biology*, eds. S. Udenfried and J. Meienhofer. Academic Press, vol. 9, ch. 3.

   (1995). In *Enzyme Catalysis in Organic Synthesis*, eds. K. Drauz and H. Waldmann. VCH vol. 1, pp. 431–58.

Jakubke, H.-D., Kuhl, P., and Könnecke, A. (1985). Basic principles of protease-catalyzed peptide bond formation. *Angew. Chem. Int. Ed. Engl.*, **24**, 85–93.

Jakubke, H.-D., Eichhorn, U., Hansler, M., and Ullmann, D. (1996). Non-conventional enzyme catalysis: application of proteases and zymogens in biotransformations. *Biol. Chem.*, **377**, 455–64.

Janiak, M. J., Small, D. M., and Shipley, G. G. (1976). Nature of the thermal pretransition of synthetic phospholipids: dimyristoyl- and dipalmitoyllecithin. *Biochemistry*, **15**, 4575–80.

Jay, D. and Gilbert, W. (1987). Basic protein enhances the encapsulation of DNA into lipid vesicles: model for the formation of primordial cells. *Proc. Natl. Acad. Sci. USA*, **84**, 1978–80.

Jimenez-Prieto, R., Silva, M., and Perez-Bendito, D. (1998). Approaching the use of oscillating reactions for analytical monitoring. *Analyst*, **123**, 1R–8R.

Joyce, G. F. (1989). RNA evolution and the origin of life. *Nature*, **338**, 217–24.

Joyce, G. F. and Orgel, L. E. (1986). Nonenzymatic template-directed synthesis on RNA random copolymers – poly(C, G) templates. *J. Mol. Biol.*, **188**, 433–41.

   (1993). Prospects for understanding the origin of the RNA World. In *The RNA World*, eds. R. F. Gesteland and J. F. Atkins. Plainview, Cold Spring Harbor Laboratory Press, pp. 1–25.

Joyce, J. (1994). In *Origins of Life: The Central Concepts*, eds. D. W. Deamer and G. R. Fleischaker. Jones and Bartlett, foreword.

Kaler, E. W., Murthy, A. K., Rodriguez, B. E., and Zasadzinski, J. A. N. (1989). Spontaneous vesicle formation in aqueous mixtures of single-tailed surfactants. *Science*, **245**, 1371–4.

Karnup, A. S., Uverskii, V. N., and Medvedkin, V. N. (1996). Synthetic Polyaminoacids and Polypeptides, Preparation by the N-carboxyanhydride method. *Russ. J. Bioorg. Chem.*, **22**, 479–90.

Kaszuba, M. and Jones, M. N. (1999). Hydrogen peroxide production from reactive liposomes encapsulating enzymes. *Biochim. Biophys. Acta*, **1419**, 221–8.

Kauffman, S. A. (1986). Autocatalytic set of proteins. *J. Theor. Biol.*, **119**, 1–24.

   (1993). *The Origins of Order*. Oxford: Oxford University Press.

Kawamura, K. and Kamoto, F. (2000). Condensation reaction of hexanucleotides containing guanine and cytosine with water soluble carbodiimide. *Nucleic Acid Symp. Ser.*, **44**, 217–18.

Kawamura, K. (2002). The origin of life from the life of subjectivity. In *Fundamentals of life*, eds. G. Palyi, C. Zucchi, and L. Caglioti. Elsevier, pp. 563–76.

Kent, S. (1999). Chemical protein synthesis by solid phase ligation of unprotected peptide segments. *J. Am. Chem. Soc.*, **121**, 8720–7.

Kessaissia, S., Siffert, B., and Donnet, J. B. (1980). Synthese de peptides; preparation de l'acide hippurique par reaction des complexes montmorillonite-glycine avec l'acide benzoique. *Clay Minerals*, **15**, 383–92.

Kiedrowski, G. von (1986). A self-replicating hexadeoxynucleotide. *Angew. Chem. Int. Ed. Engl.*, **25**, 932–5.

  (1993). Minimal replicator theory I: parabolic versus exponential growth. In *Bioorganic Chemistry*, ed. D. H. Berlin. Springer Verlag, vol. 3, pp. 115–46.

Kikuchi, A., Aoki, Y., Sugaya, S., *et al.* (1999). Development of novel cationic liposomes for efficient gene transfer into peritoneal disseminated tumor. *Human. Gene Ther.* **10** (6), 947–55.

Kim, J. (1984). Concepts of supervenience. *Phil. Phen. Res.*, **45**, 153–76.

Kimura, M. (1983). *The Neutral Theory of Molecular Evolution*. Cambridge University Press.

Kirstein, S., von Berlepsch, H., Bottecher, C., *et al.* (2000). Chiral J-aggregates formed by achiral cyanine dyes. *Chem. Phys. Chem.*, **1**, 146–50.

Klee, R. (1984). Micro-determinism and concepts of emergence. *Phil. Sci.*, **51**, 44–63.

Knenvolden, K., Lawless, J. G., Pering, K., *et al.* (1970). Evidence for extraterrestrial amino acids and hydrocarbons in the Murchison meteorite. *Nature*, **228**, 923–6.

Kolisnychenko, V., Plunkett, G., III, Herring, C. D., *et al.* (2002). Engineering a reduced *Escherichia coli* genome. *Genome Res.*, **12**, 640–7.

Kondepudi, D. K. and Prigogine, I. (1981). Sensitivity of non-equilibrium systems. *Physica A*, **107**, 1–24.

Kondepudi, D. K., Kaufman, R., and Singh, N. (1990). Chiral symmetry breaking in sodium chlorate crystallization. *Science*, **250**, 975.

Kondepudi, D. K., Prigogine, I., and Nelson, G. (1985). Sensitivity of branch selection in nonequilibrium systems. *Phys. Lett. A*, **111**, 29–32.

Kondo, Y., Uchiyama, H., Yoshino, N., Nishiyama, K., and Abe, M. (1995). Spontaneous vesicle formation from aqueous-solutions of didodecyldimethylammonium bromide and sodium dodecyl-sulfate mixtures. *Langmuir*, **11**, 2380–4.

Koonin, E. V. (2000). How many genes can make a cell: the minimal-gene-set concept. *Annu. Rev. Genomics Human Genet.* **1**, 99–116.

Koshland, D. E., Jr. (2002). The seven pillars of life. *Science.* **295**, 2215–16.

Koster, G., Van Duijin, M., Hofs, B., and Dogterom, M. (2003). Membrane tube formation from giant vesicles by dynamic association of motor proteins. *Proc. Natl. Acad. Sci.*, **100**, 15583–8.

Kricheldorf, H. R. (1990). In *Models of Biopolymers by Ring-Opening Polymerization*, ed. S. Penczek. CRC Press, pp. 46–62.

Kricheldorf, H. R. and Hull, W. E. (1979). Stereospecificity of the polymerization of D, L-alanine-NCA and D,L-alanine NCA. *Makromol. Chem.*, **180**, 1715–24.

Kuiper, T. B. H. and Morris, M. (1977). Searching for extraterrestial civilizations. *Science*, **196**, 616–21.

Kullmann, W. (1987). *Enzymatic Peptide Synthesis*. CRC Press.

Lahav, M. and Leiserowitz, L. (1999). Spontaneous resolution: from three-dimensional crystals to two-dimensional magic nanoclusters. *Angew. Chem. Int. Ed. Engl.*, **38**, 2533–6.

Landau, E. M. and Luisi, P. L. (1993). Lipid cubic phases as transparent, rigid matrices for the direct spectroscopic study of immobilized membrane proteins. *J. Am. Chem. Soc.*, **115**, 2102–6.

Landweber, L. F. and Pokrovskaya, I. D. (1999). Emergence of a dual-catalytic RNA with metal-specific cleavage and ligase activities: the spandrels of RNA evolution. *Proc. Natl. Acad. Sci.*, **96**, 173–8.

Langton, C. G. (1990). Computation at the edge of chaos: phase transitions and emergent computation. *Physica*, **D42**, 12–37.

Larsson, K. (1989). Cubic lipid-water phases: structures and biomembrane aspects. *J. Phys. Chem.*, **93**, 7304–14.

Lasch, J., Laub, R., and Wohlrab, W. (1991). How deep do intact liposomes penetrate into human skin? *J. Controll. Release*, **18**, 55–8.

Lasic, D. D. (1995). In *Handbook of Biological Physics*, eds. R. Lipowsky and E. Sackmann. Elsevier vol. 1, pp. 491–519.

Lauceri, R., Raudino, A., Scolaro, L. M., Mical, N., and Purrello, R. (2002). From achiral porphyrins to template-imprinted chiral aggregates and further: self-replication of chiral memory from scratch. *J. Am. Chem. Soc.*, **124**, 894–5.

Lawless, J. G. and Yuen, G. U. (1979). Quantification of monocarboxylic acids in the Murchison carbonaceous meteorite. *Nature*, **282**, 396–8.

Lawrence, D. S., Jiang, T., and Levett, M. (1995). Self-assembling supramolecular complexes. *Chem. Rev.*, **95**, 2229–60.

Lazcano, A. (2003). Just how pregnant is the universe? *Science*, **299**, 347–8.

Lazcano, A. (2004). An answer in search of a question how life began: the genesis of life on Earth, by William Day. *Astrobiology*, **4**, 469–71.

Lazcano, A. (2005). Teaching evolution in Mexico: preaching to the choir. *Science*, **310**, 787–9.

Lazcano, A. and Bada, J. L. (2003). The 1953 Stanley L. Miller experiment: fifty years of prebiotic organic chemistry. *Orig. Life Evol. Biosph.*, **33**, 235–42.

Lazcano, A., Guerriero, R., Margulius, L., and Oró, J. (1988). The evolutionary transition from RNA to DNA in early cells. *J. Mol. Evol.*, **27**, 283–90.

Lazcano, A., Valverde, V., Hernandez, G., *et al.* (1992). On the early emergence of reverse transcription: theoretical basis and experimental evidence. *J. Mol. Evol.*, **35**, 524–36.

Lazzara, S. (2001). *Vedi Alla Voce Scienza*. Manifesto Libri.

le Doux, J. (2002). *Synaptic Self : How Our Brains Become Who We Are*. Viking Books.

Lee, D. H., Granja, J. R., Martinez, J. A., Severin, K., and Ghadiri, M. R. (1996). A self-replicating peptide. *Nature*, **382**, 525–8.

Lee, D. H., Severin, K., Yokobayashi, Y., and Ghadiri, M. R. (1997). Emergence of symbiosis in peptide self-replication through a hypercyclic network. *Nature*, **390**, 591–4.

Leman, L., Orgel, L., and Ghadiri, M. R. (2004). Carbonyl sulfide-mediated prebiotic formation of peptides. *Science*, **306**, 283–6.

Leser, M. E. and Luisi, P. L. (1989). Liquid 3-phase micellar extraction of peptides. *Biotech. Techniques*, **3**, 149–54.

(1990). Application of reverse micelles for the extraction of amino acids and proteins. *Chimia*, **44**, 270–82.

Levashov, A. V., Klyachko, N. L., Psbezhetski, A. V., *et al.* (1989). *Biochim. Biophys. Acta*, **988**, 221–56.

Levy, M. and Ellington, A. D. (2003). Peptide-template nucleic acid ligation, *J. Mol. Evol.*, **56**, 607–15.

Lewontin, R. C. (1993). *The Doctrine of DNA – Biology as an Ideology*. Penguin Books.

Li, T. and Nicolaou, K. C. (1994). Chemical self-replication of palindromic duplex DNA. *Nature*, **369**, 218–21.

Li, Y., Zhao, Y., Hatfield, S. *et al.* (2000). Dipeptide seryl-histidine and related oligopeptides cleave DNA, protein, and a carboxyl ester. *Bioorg. Med. Chem.*, **12**, 2675–80.

Li, X. and Chmielewski, J. (2003). Peptide self-replication enhanced by a proline kink. *J. Am. Chem. Soc.*, **125**, 11820–1.

Lifson, S. (1997). On the crucial stages in the origin of animate matter. *J. Mol. Evol.*, **44**, 1–8.

Limtrakul, J., Kokpol, S., and Rode, B. M. (1985). Quantum chemical investigation on ion-dipeptide complex-formation. *J. Sci. Soc. Thailand*, **11**, 129–33.

Lindblom, G. and Rilfors, L. (1989). Cubic phases and isotropic structures formed by membrane lipids – possible biological relevance. *Biochem. Biophys. Acta*, **988**, 221–56.

Lindsey, J. S. (1991). Self-assembly in synthetic routes to molecular devices – biological principles and chemical perspectives – a review. *New J. Chem.*, **15**, 153–80.

Livio, M. and J. Rees, M. (2005). Anthropic reasonings. *Science*, **309**, 1922–3.

Lonchin, S., Luisi, P. L., Walde, P., and Robinson, B. H. (1999). A matrix effect in mixed phospholipid/fatty acid vesicle formation. *J. Phys. Chem. B*, **103**, 10910–16.

Love, S. G. and Brownlee, D. E. (1993). A direct measurement of the terrestrial mass accretion rate of cosmic dust. *Science*, **262**, 550–3.

Lovelock, J. E. (1979). *Gaia: A New Look at Life on Earth*. Oxford University Press.
(1988). *The Ages of Gaia*. W. W. Norton & Co.

Luci, P. (2003). Gene cloning expression and purification of membrane proteins. ETH-Z Dissertation Nr. 15108, Zurich.

Luhmann, K. (1984). *Soziale Systeme*. Suhrkamp.

Luisi, P. L. (1979). Why are enzymes macromolecules? *Naturwissenschaften*, **66**, 498–504.
(1985). Enzyme hosted in reverse micelles in hydrocarbon solution. *Angew. Chem.*, **24**, 439–50.
(1993). Defining the transition to life: self-replicating bounded structures and chemical autopoiesis. In *Thinking About Biology*, SFI Studies in the Sciences of Complexity, ed. W. Stein and F. J. Varela. Addison-Wesley-Longman.

Luisi, P. L. (1994). In *Self-Reproduction of Supramolecular Structures*, *Proceedings from the Maratea Symposium*, eds. G. Fleischacker, S. Colonna and P. L. Luisi. Kluwer.
(1996). Self-reproduction of micelles and vesicles: models for the mechanisms of life from the perspective of compartmented chemistry. *Adv. Chem. Phys.*, **92**, 425–38.
(1997a). Die Frage nach dem Ursprung des Lebens. *Schweiz. Techn. Z.*, **12**, 10–14.
(1997b). Self-reproduction of chemical structures and the question of the transition to life. In *Astronomical and Biochemical Origins and the Search for Life in the Universe*, eds. C. B. Cosmovici, S. Bowyer, and D. Werthimer. Editrice Compositori, pp. 461–8.

Luisi, P. L. and Laane, C. (1986). Solubilization of enzymes in apolar solvents via reverse micelles. *Trends Biotechnol.*, **4**, 29–38.
(1998). About various definitions of life. *Orig. Life Evol. Biosph.*, **28**, 613–22.
(2001). Are micelles and vesicles chemical equilibrium systems? *J. Chem. Educ.*, **78**, 380–4.
(2002a). Toward the engineering of minimal living cells. *Anat. Rec.*, **268**, 208–14.
(2002b). Emergence in chemistry: chemistry as the embodiment of emergence. *Found. Chem.*, **4**, 183–200.
(2003a). Contingency and determinism. *Phil. Trans. R. Soc. Lond., A*, **361**, 1141–7.
(2003b). Autopoiesis: a review and reappraisal. *Naturwissenschaften*, **90**, 49–59.

Luisi, P. L. and Magid, L. (1986). Solubilization of enzymes and nucleic acids in hydrocarbon micelar solutions. *Crit. Rev. Biochem.*, **20**, 409–74.

Luisi, P. L. and Oberholzer, T. (2001). Origin of life on Earth: molecular biology in liposomes as an approach to the minimal cell. In *The Bridge between the Big Bang and Biology*, ed. F. Giovanelli. CNR Press, pp. 345–55.

Luisi, P. L. and Straub, B., eds. (1984). *Reverse Micelles*. Plenum Press.

Luisi, P. L. and Varela, F. J. (1990). Self-replicating micelles – a chemical version of minimal autopoietic systems. *Orig. Life Evol. Biosph.*, **19**, 633–43.

Luisi, P. L. and Walde, P., eds. (2000). *Giant Vesicles, Perspectives in Supramolecular Chemistry*. John Wiley & Sons Ltd.

Luisi, P. L., Ferri, F., and Stano, P. (2006). Approaches to semi-synthetic minimal cells: a review. *Naturwissenschaften*, **93**, 1–13.

Luisi, P. L., Lazcano, A., and Varela, F. (1996). In *Defining Life: the Central Problem in Theoretical Biology*, ed. M. Rizzotti. University of Padova, pp. 149–65.

Luisi, P. L., Oberholzer, T., and Lazcano, A. (2002). The notion of a DNA minimal cell: a general discourse and some guidelines for an experimental approach. *Helv. Chim. Acta*, **85** (6), 1759–77.

Luisi, P. L., Pellegrini, A., and Walsoe, C. (1977b). Pepsin-catalyzed coupling between aromatic amino acid residues. *Experientia*, **33**, 796.

Luisi, P. L., Giomini, M., Pileni, M. P., and Robinson, B. H. (1988). Reverse micelles as hosts for proteins and small molecules. *Biochim. Biophys. Acta*, **947**, 209–46.

Luisi, P. L., Scartazzini, R., Haering, G., and Schurtenberger, P. (1990). Organogels from water-in-oil microemulsions. *Colloid Polymer Sci.*, **268**, 356–74.

Luisi, P. L., Stano, P., Rasi, S., and Mavelli, F. (2004). A possibile route to prebiotic vesicle reproduction, *Artificial Life*, **10**, 297–308.

Luisi, P. L., Bonner, F. J., Pellegrini, A., Wiget, P., and Wolf, R. (1979). Micellar solubilization of proteins in aprotic solvents and their spectroscopic properties. *Helv. Chim. Acta*, **62**, 740–53.

Luisi, P. L., Henninger, F., Joppich, M., Dossena, A., and Casnati, G. (1977a). Solubilization and spectroscopic properties of α-chymotrypsin in cyclohexane. *Biochem. Biophys. Res. Commun.*, **74**, 1384–9.

Luther, A., Brandsch, R., and von Kiedrowski, G. (1998). Surface promoted replication and exponential amplification. *Nature*, **396**, 245–8.

Luthi, P. and Luisi, P. L. (1984). Enzymatic-synthesis of hydrocarbon-soluble peptides with reverse micelles. *J. Am. Chem. Soc.*, **106**, 7285–6.

Luzzati, V., Vargas, R., Mariani, P., Gulik, A., and Delacroix, H. (1993). Cubic phases of lipid-containing systems: elements of a theory and biological connotations. *J. Mol. Biol.*, **229**, 540–51.

Ma, Q. G. and Remsen, E. F. (2002). Chemically induced supramolecular reorganization of triblock copolymer assemblies: Trapping of intermediate states via a shell-crosslinking methodology. *Proc. Natl. Acad. Sci. USA*, **99**, 5058–63.

Machy, P. and Leserman, L. (1987). *Liposomes in Cell Biology and Pharmacology*. London: John Libbey and Co. Ltd.

Madeira, V. M. C. (1977). *Biochim. Biophys. Acta*, **499**, 202–211.

Mader, S. S. (1996). *Biology*, 5th edn. W. C. Brown Publisher.

Maestro, M. and Luisi, P. L. (1990). A simplified thermodynamic model for protein uptake by reverse micelles. In *Surfactants in Solution*, ed. K. L. Mittal. Plenum, vol. 9.

Margulis, L. (1993). *Symbiosis in Cell Evolution*. Freeman.

Margulis, L. and Sagan, D. (1995). *What is Life?* Weidenfeld and Nicholson.

Mariani, P., Luzzati, V., and Delacroix, H. (1988). Cubic phases of lipid-containing systems: structure analysis and biological implications. *J. Mol. Biol.*, **204**, 165–89.

Marks-Tarlow, T., Robertson, R., and Combs, A. (2001). Varela and the Uroborus: the psychological significance of reentry. *Cybernetics Human Knowing*, **9**, 31.

Marques, E. F., Regev, O., Khan, A., Miguel, M. D., and Lindman, B. (1998). Vesicle formation and general phase behavior in the catanionic mixture SDS-DDAB-water. The anionic-rich side. *J. Phys. Chem. B*, **102**, 6746–58.

Martinek, K. and Berezin, I. V. (1986). *Dokl. Akdam. Nauk., SSSR*, **289**, 1271.

Martinek, K., Levashov, A. V., Pantin, V. I., and Berezin, I. V. (1978). Model of biological membranes or surface-layer (active center) of protein globules (enzymes) – reactivity of water solubilized by reversed micelles of aerosol OT in octane during neutral hydrolysis of picrylchloride. *Doklady Akademii Nauk SSSR*, **238**, 626–9.

Martinek, K., Levashov, A. V., Klyachko, N. L., Pantin, V. I., and Berezin, I. V. (1981). The principles of enzyme stabilization. 6. Catalysis by water-soluble enzymes entrapped into reversed micelles of surfactants in organic solvents. *Biochem. Biophys. Acta*, **657**, 277–95.

Martinek, K, Levashov, A. V., Klyachko, N., Khmelnttski, Y. L., and Berezin, I. V. (1986a). Micellar enzymology. *Eur. J. Biochem.*, **155**, 453–468.

Mason, S. F. and Tranter, G. E. (1983). The parity violating energy difference between enantiomeric molecules. *Chem. Phys. Lett.*, **94**, 34.

(1984). The parity violating energy difference between enantiomeric molecules. *Mol. Phys.*, **53**, 1091–111.

Matsumura, S., Takahashi, T., Ueno, A., and Mihara, H. (2003). Complementary nucleobase interaction enhances peptide–peptide recognition and self-replicating catalysis. *Chem. Eur. J.*, **9**, 4829–37.

Matthews, C. N. (1975). The origin of proteins, heteropolypeptides from hydrogen cyanide and water, *Origin of Life*, **6**, 155–63.

Maturana, H. and Varela, F. (1980). *Autopoiesis and Cognition: The Realization of the Living*. Reidel.

(1998). *The Tree of Knowledge*, revised edn. Shambala.

Maturana, H., Lettvin, J., McCulloch, W., and Pitts, W. (1960). Life and cognition. *Gen. Physiol.*, **43**, 129–75.

Mavelli, F. (2004). Theoretical investigations on autopoietic replication mechanisms. ETH-Z Dissertation Nr. 15218, Zurich.

Mavelli, F. and Luisi, P. L. (1996). Autopoietic self-reproducing vesicles: a simplified kinetic model. *J. Phys. Chem.*, **100**, 16600–7.

Maynard-Smith, J. and Szathmáry, E. (1995). *The Major Transitions in Evolution*. Oxford University Press.

(1999). *The Origins of Life*. Oxford.

Mayr, E. (1988). The limits of reductionism. *Nature*, **331**, 475.

McBride, J. M. and Carter, R. L. (1991). Spontaneous resolution by stirred crystallization. *Angew. Chem. Int. Ed. Engl.*, **30**, 293–5.

McCollom, T. M., Ritter, G., and Simoneit, B. R. T. (1999). Lipid synthesis under hydrothermal conditions by Fischer-Tropsch-type reactions. *Orig. Life Evol. Biosph.*, **29**, 153–66.

McLaughlin, B. P. (1992). The rise and fall of British emergentism. In *Emergence or Reduction: Essays on the Prospects of Nonreductive Materialism*, eds. A. Beckermann, H. Flohr and J. Kim. de Gruyter, pp. 49–3.

Meier, C. A. (1992). *Wolfang Pauli und C. G. Jung, Ein Briefwechsel*. Springer Verlag.

Menger, F. (1991). Groups of organic molecules that operate collectively. *Angew. Chem. Int. Ed. Engl.*, **30**, 1086–99.

Merleau-Ponty, M. (1967). *The Structure of Behaviour*. Beacon.

Micura, R., Bolli, M., Windhab, N., and Eschenmoser, A. (1997). *Angew. Chem. Int. Ed. Engl.*, **36**, 870.

Micura, R., Kudick, R., Pitsch, S., and Eschenmoser, A. (1999). *Angew. Chem. Int. Ed. Engl.*, **38**, 680.

Mill, J. S. (1872). *System of Logic*, 8th edn. Longmans, Green, Reader and Dyer.

Miller, C., Cuendet, P., and Gratzel, M. (1991). Adsorbed omega-hydroxy thiol monolayers on gold electrodes – evidence for electron-tunneling to redox species in solution. *J. Phys. Chem.*, **95**, 877–86.

Miller, M. B. and Basler, B. L. (2001). Quorum sensing in bacteria. *Ann. Rev. Microbiol.*, **55**, 165–199.

Miller, S. L. (1953). Production of amino acids under possible primitive Earth conditions. *Science*, **117**, 2351–61.

(1998). The endogenous synthesis of organic compounds. In *The Molecular Origin of Life*, ed. A. Brack. Cambridge University Press.

Miller, S. L. and Bada, J. (1988). Submarine hot springs and the origin of life. *Nature*, **334**, 609–11.

(1991). Extraterrestrial synthesis. *Nature*, **350**, 388–89.

Miller, S. L. and Lazcano, A. (1995). *J. Mol. Evol.*, **41**, 689–92.

(2002). Formation of the building blocks of life. In *Life's Origin, The Beginning of Biological Evolution*, ed. J. W. Schopf. California University Press, pp. 100–9.

Miller, S. L. and Parris, M. (1964). *Nature*, **204**, 1248–50.

Mingers, J. (1992). The problems of social autopoiesis. *Int. J. Gen. Syst.*, **21**, 229–36.

(1995). *Self-Producing Systems: Implications and Applications of Autopoiesis*. Plenum.

(1997). A critical evaluation of Maturana's constructivist family therapy. *Syst. Practice*, **10** (2), 137–51.

Miranda, M., Amicarelli, F., Poma, A., Ragnelli, A. M., and Arcadi, A. (1988). *Biochim. Biophys. Acta*, **966**, 276–86.

Monod, J. (1971). *Chance and Necessity*. A. A. Knopf.

Morgan, C. L. (1923). *Emergent Evolution*. William and Norgate.

Morigaki, K., Dallavalle, S., Walde, P., Colonna, S., and Luisi, P. L. (1997). Autopoietic self-reproduction of chiral fatty acid vesicles. *J. Am. Chem. Soc.*, **119**, 292–301.

Morowitz, H. J. (1967). Biological self-replicating systems. *Prog. Theor. Biol.*, **1**, 35–58.

(1992). *Beginnings of Cellular Life*. Yale University Press.

Morowitz, H. J., Deamer, D. W., and Smith, T. (1991). Biogenesis as an evolutionary process. *J. Mol. Evol.*, **33**, 207–8.

Morowitz, H. J., Peterson, E., and Chang, S. (1995). The synthesis of glutamic acid in the absence of enzymes – implications for biogenesis. *Orig. Life Evol. Biosph.*, **25**, 395–9.

Morowitz, H. J., Kostelnik, J. D., Yang, J., and Cody, G. D. (2000). The origin of intermediary metabolism. *Proc. Natl. Acad. Sci. USA*, **97**, 7704–9.

Mossa, G., Di Giulio, A., Dini, L., and Finazzi-Agrò, A. (1989). *Biochim. Biophys. Acta*, **986**, 310–14.

Müller, D., von Pitch, S., and Kittaka, A. (1990). Chemie von α-aminonitriles, *Helv. Chim. Acta*, **73**, 1410–68.

Mushegian, A. (1999). The minimal genome concept. *Curr. Opin. Genetics Develop.*, **9**, 709–714.

Mushegian, A. and Koonin, E. V. (1996). A minimal gene set for cellular life derived by comparison of complete bacterial genomes. *Proc. Natl. Acad. Sci. USA*, **93**, 10268–73.

Nagel, E. (1961). *The Structure of Science*. Harcourt.

Nakajima, T., Yabushita, Y., and Tabushi, I. (1975). Amino acid synthesis through biogenic CO_2 fixation. *Nature*, **256**, 60–1.

Naoi, M., Naoi, M., Shimizu, T., Malviya, A. N., and Yagi, K. (1977). Permeability of amino acids into liposomes. *Biochim. Biophys. Acta*, **471**, 305–10.

Neumann, J. von and Burks, A., eds. (1966). *Theory of Self-Reproduction Automata*. University of Illinois Press.

Nicolis, G. and Prigogine, I. (1977). *Self-Organization in Nonequilibrium Systems. From Dissipative Structures to Order Through Fluctuations*. Wiley.

Nissen, P., Hansen, J., Ban, N., Moore, P. B., and Steitz, T. A. (2000). The structural basis of ribosome activity in peptide bond synthesis. *Science*, **289**, 920–30.

Noireaux, V. and Libchaber, A. (2004). A vesicle bioreactor as a step toward an artificial cell assembly. *Proc. Natl. Acad. Sci. USA*, **101**, 17669–74.

Noireaux, V., Bar-Ziv, R., and Libchaber, A. (2003). Principles of cell-free genetic circuit assembly. *Proc. Natl. Acad. Sci. USA*, **100**, 12672–7.

Nomura, S. M., Tsumoto, K., Yoshikawa, K., Ourisson, G., and Nakatani, Y. (2002). Towards proto-cells: "primitive" lipid vesicles encapsulating giant DNA and its histone complex, *Cell. Mol. Biol. Lett.*, **7**, 245–6.

Nomura, S. M., Tsumoto, K., Hamada, T., *et al.* (2003). Gene expression within cell-sized lipid vesicles, *Chem. Bio. Chem.*, **4**, 1172–5.

Nooner, D. W., Gilbert, J. M., Gelpi, E., and Oró, J. (1976). Closed system Fischer-Tropsch synthesis over meteoritic iron, iron-ore and nickel-iron alloy. *Geochim. Cosmochim. Acta*, **40**, 915–24.

Noyes, R. M. (1989). Some models of chemical oscillators. *J. Chem. Educ.*, **66**, 190–1.

O'Connor, T. (1994). Emergent properties. *Am. Phil. Q.*, **31**, 91–104.

Oberholzer, T. and Luisi, P. L. (2002). The use of lipsomes for constructing cell models. *J. Biol. Phys.*, **28**, 733–44.

Oberholzer, T., Albrizio, M., and Luisi, P. L. (1995a). Polymerase chain reaction in liposomes. *Curr. Biol.*, **2**, 677–82.

Oberholzer, T., Nierhaus, K. H., and Luisi, P. L. (1999). Protein expression in liposomes. *Biochem. Biophys. Res. Commun.*, **261**, 238–41.

Oberholzer, T., Wick, R., Luisi, P. L., and Biebricher, C. K. (1995b). Enzymatic RNA replication in self-reproducing vesicles: an approach to a minimal cell. *Biochem. Biophys. Res. Commun.*, **207**, 250–7.

Oie, T., Loew, G. H., Burt, S. K., and MacElroy, R. D. (1983). Quantum chemical studies of a model for peptide bond formation. 2. Role of amine catalyst in formation of formamide and water from ammonia and formic acid. *J. Am. Chem. Soc.*, **105**, 2221–7.

Olsson, U. and Wennerstrom, H. (2002). On the ripening of vesicle dispersions. *J. Phys. Chem. B*, **106**, 5135–8.

Ono, N. and Ikegami, T. (2000). Self-maintenance and self-reproduction in an abstract cell model. *J. Theor. Biol.*, **206**, 243–53.

Oparin, A. I. (1924). *Proishkhozhddenie Zhisni*. Moskowski Rabocii.
  (1938). *Origin of Life*. McMillan.
  (1953). *The Origin of Life*. Dover Publications.
  (1957). *The Origin of Life on Earth*, 3rd edn. Academic Press.
  (1961). *Life: Its Nature, Origin and Development*. Oliver and Boyd.

Oppenheim, P. and Putnam, H. (1958). The unity of science as a working hypothesis. In *Minnesota Studies in the Philosphy of Science*, eds. H. Feigl, G. Maxwell and M. Scriven. University of Minnesota Press, pp. 3–36.

Orgel, L. E. (1973). *The Origins of Life*. Wiley.

Orgel, L. E. (1992). Molecular replication. *Nature*, **358**, 203–9.

(1994). The origin of life on the Earth. *Sci. Amer.*, **271** (4), 53–61.

(1995). Unnatural selection in chemical systems. *Acc. Chem. Res.*, **28**, 109–18.

(1998). Polymerization on the rocks: theoretical introduction. *Orig. Life Evol. Biosph.*, **28**, 227–34.

(2000). Self-organizing biochemical cycles. *Proc. Natl. Acad. Sci. USA*, **97**, 12503–7.

(2002). The origin of biological information. In *Life's Origin, the Beginnings of Biological Evolution*, ed. J. W. Schopf. California University Press, pp. 1–38.

(2003). Some consequences of the RNA world hypothesis. *Orig. Life Evol. Biosph.*, **33**, 211–18.

Oró, J. (1960). Synthesis of adenine from ammonium cyanide. *Biochem. Bioph. Res. Commun.*, **2**, 407–12.

(1961). Amino acid synthesis from hydrogen cyanide under possible primitive Earth conditions. *Nature*, **190**, 442–3.

(1994). In *Early Life on Earth, Nobel Symposium n. 84*, ed. S. Bengtson. Columbia University Press, pp. 48–59.

(2002). Historical understanding of life's origin. In *Life's Origin, the Beginnings of Biological Evolution*, ed. J. W. Schopf. California University Press, pp. 7–41.

Oró, J. and Kimball, A. P. (1961). Synthesis of purines under possible primitive Earth conditions. 1. Adenine from hydrogen cyanide. *Arch. Biochem. Biophys.*, **94**, 221–7.

Oró, J. and Kimball, A. P. (1962). Synthesis of purines under possible primitive earth conditions. 2. Purine intermediates from hydrogen cyanide. *Arch. Biochem. Biophys.*, **96**, 293–313.

Ousfouri, S., Stano, P., and Luisi, P. L. (2005). Condensed DNA in lipid microcompartments. *J. Phys. Chem. B.*, **109**, 19929–35.

Ourisson, G. and Nakatani, Y. (1994). The terpenoid theory of the origin of cellular life: the evolution of terpenoids to cholesterol. *Chem. Biol.*, **1**, 11–23.

(1999). Origin of cellular life: molecular foundations and new approaches. *Tetrahedron*, **55**, 3183–90.

Paechthorowitz, M. and Eirich, F. R. (1988). The polymerisation of amino acid adenilates on sodium montmorillonite with preadsorbed polypeptides. *Orig. Life Evol. Biosph.*, **18**, 359–87.

Palazzo, G. and Luisi, P. L. (1992). Solubilization of ribosomes in reverse micelles. *Biochem. Biophys. Res. Commun.*, **186**, 1546–52.

Paley, W. (1802) (other sources report 1803). *Natural Theology, or Evidences of the Existence and Attributes of the Deity, Collected from the Appearances of Nature*, 12th edn (1986). Lincoln-Rembrandt Publishing.

Palyi G., Zucchi C., and Caglioti L., eds. (2002). *Fundamentals of life*. Elsevier.

Pantazatos, D. P. and McDonald R. C. (1999). Directly observed membrane fusion between oppositely charged phospholipid bilayers. *Membrane Biol.*, **170**, 27–38.

Papahadjopoulus, D., Lopez, N., and Gabizon, A. (1989). In *Liposomes in Therapy of Infectious Diseases and Cancer*, eds. G. Lopez-Berenstein and I. J. Fidler. Alan Riss Inc., pp. 135–154.

Parsons, P. (1996). Dusting off panspermia. *Nature*, **383**, 221–2.

Paul, N. and Joyce, G. F. (2002). A self-replicating ligase ribozyme. *Proc. Natl. Acad. Sci.*, **99**, 12733–40.

(2004). Minimal self-replicating systems. *Curr. Opin. Chem. Biol.*, **8**, 634–9.

Penzien, K. and G. M. J. Schmidt (1969). Reactions in chiral crystals – an absolute asymmetric synthesis. *Angew. Chem. Int. Ed. Engl.*, **8**, 608.

Peretó, Y., Lopez-Garcia, P., and Moreira, D. (2004). Ancestral lipid biosynthesis and early membrane evolution. *Trends Biochem. Sci.*, **29**, 469–77.

Pfammatter, N., Guadalupe, A. A., and Luisi, P. L. (1989). Solubilization and activity of yeast cells in water-in-oil microemulsion. *Biochem. Biophys. Res. Commun.*, **161**, 1244–51.

Pfammatter, N., Hochkoppler, A., and Luisi, P. L. (1992). Solubilization and growth of *Candida pseudotropicalis* in water-in-oil microemulsions. *Biotechnol. Bioeng.*, **40**, 167–72.

Pfüller, U. (1986). *Mizellen, Vesikeln, Mikroemulsionen*. Springer Verlag.

Piaget, J. (1967). *Biologie et connaissance*. Gallimard.

Pietrini, A. V. and Luisi, P. L. (2002). Circular dichroic properties and average dimensions of DNA-containing reverse micellar aggregates. *Biochim. Biophys. Acta*, **1562**, 57–62.

(2004). Cell-free protein synthesis through solubilisate exchange in water/oil emulsion compartments. *Chem. Bio. Chem*, **5**, 1055–1062.

Pileni, M. P. (1981). Photoelectron transfer in reverse micelles – photo-reduction of cytochrome-c. *Chem. Phys. Lett.*, **81**, 603–5.

Piries, N. M. (1953). *Discovery*, **14**, 238.

Pizzarello, S. and Cronin, J. R. (2000). Non-racemic amino acids in the Murray and Murchison meteorites. *Geochim. Cosmochim. Acta*, **64**, 329–38.

Pizzarello, S. and Weber, A. L. (2004). Prebiotic amino acids as asymmetric catalysts. *Science*, **303**, 1151.

Pizzarello, S., Feng, X., Epstein, S., and Cronin, J. R. (1994). Isotopic N-lyses of nitrogenous compounds from Murchison meteorite: ammonia, amines, amino acids, and polar hydrocarbons. *Geochim. Cosmochim. Acta*, **58**, 5579–87.

Plankensteiner, K., Righi, A., and Rode, B. M. (2002). Glycine and diglycine as possible catalytic factors in the prebiotic evolution of peptides. *Orig. Life Evol. Biosph.*, **32**, 225–36.

Plasson, R., Biron, J. P., Cottet, H., Commeyras, A., and Taillades, J. (2002). Kinetic study of the polymerization of alpha-amino acid N-carboxyanhydrides in aqueous solution using capillary electrophoresis. *J. Chromatogr. A*, **952**, 239–48.

Platt, J. R. (1961). Properties of large molecules that go beyond the properties of their chemical sub-groups. *J. Theor. Biol.*, **1**, 342–58.

Poerksen, B. (2004). *The Certainty of Uncertainty, Dialogues Introducing Constructivism*. Imprint Academic.

Pohorille, A. and Deamer, D. (2002). Artificial cells: prospects for biotechnology. *Trends Biotech.*, **20**, 123–8.

Pojman, J. A., Craven, R., and Leard, D. C. (1994). Oscillations and chemical waves in the physical chemistry lab. *J. Chem. Educ.*, **71**, 84–90.

Ponce de Leon, S. and Lazcano, A. (2003). Panspermia – true or false? *Lancet*, **362**, 406–7.

Popa, R. (2004). *Between Necessity and Probability: Searching for the Definition and Origin of Life*. Springer Verlag.

Pope, M. T. and Muller, A. (1991). Polyoxometalate chemistry – an old field with new dimensions in several disciplines. *Angew. Chem. Int. Ed. Engl.*, **30**, 34–48.

Portmann, M., Landau, E. M., and Luisi, P. L. (1991). Spectroscopic and rheological studies of enzymes in rigid lipidic matrices: the case of α-chymotrypsin in a lysolecithin/water cubic phase. *J. Phys. Chem.*, **95**, 8437–40.

Pozzi, G., Birault, V., Werner, B. (1996). Single-chain polyprenyl phosphates form primitive membranes. *Angew. Chem. Int. Ed. Engl.*, **35**, 177–9.

Prigogine, I. (1997). *The End of Certainty-Time, Chaos and the New Laws of Nature.* Free Press.

Prigogine, I. and Lefever, R. (1968). Symmetry breaking instabilities in dissipative systems. *J. Chem. Phys.*, **48**, 1695–700.

Primas, H. (1985). Can chemistry be reduced in physics? *Chem. Uns. Zeit*, **19**, 160.

(1993). In *Neue Horizonte 92/93: Ein Forum der Naturwissenschaften*, ed. E. P. Fischer. München: Piper.

(1998). Emergence in exact natural sciences. *Acta Politechnica Scand.* **91**, 86–7.

Pryer, W. (1880). *Die Hypothesen über den Ursprung des Lebens.* Berlin.

Purrello, R. (2003). Lasting chiral memory. *Nature Mater.*, **2**, 216–17.

Pyun, J., Zhou, X.-Z., Drockenmuller, E., and Hawker, C. J. (2003). *Mater. Chem.*, **13**, 2653.

Quack, M. (2002). *Angew. Chem.*, **41**, 4618–30.

Quack, M. and Stohner, J. (2003a). Combined multidimensional anharmonic and parity violating effects in CDBrClF. *J. Chem. Phys.*, **119**, 11228–40.

(2003b). Molecular chirality and the fundamental symmetries of physics: influence of parity violation on rotovibrational frequencies and thermodynamic properties. *Chirality*, **15**, 375–6.

Raab, W. (1988). *Ärtzliche Kosmetologie*, **18**, 213–24.

Radzicka, A. and Wolfenden, R. (1996). Rates of uncatalyzed peptide bond hydrolysis in neutral solution and the transition state affinities of proteases. *J. Am. Chem. Soc.*, **118**, 6105–9.

Ramundo-Orlando, A., Arcovito, C., Palombo, A., Serafino, A. L., and Mossa, G. (1993). *J. Liposome Res.*, **3**, 717–24.

Ramundo-Orlando, A., Mattia, F., Palombo, A., and D'Inzeo, G. (2000). Effect of low frequency, low amplitude magnetic fields on the permeability of cationic liposomes entrapping carbonic anhydrase, part II. *Bioelectromagnetics*, **21**, 499–507.

Rao, M., Eichberg, J., and Oró, J. (1987). Synthesis of phosphatidylethanolamine under possible primitive earth conditions. *J. Mol. Evol.*, **25**, 1–6.

Rasi, S., Mavelli, F., and Luisi, P. L. (2003). Cooperative micelle binding and matrix effect in oleate vesicle formation. *J. Phys. Chem. B*, **107**, 14068–76.

(2004). Matrix effect in oleat-micelles-vesicles transformations. *Orig. Life Evol. Bioph.*, **34**, 215–24.

Rathman, J. F. (1996). Micellar catalysis. *Curr. Opin. Coll. Interf. Sci.*, **1**, 514–518.

Rebek, J. (1994). A template for life. *Chem. Br.*, **30**, 286–90.

Reichenbach, H. (1978). The aims and methods of physical knowledge. In *Hans Reichenbach: Selected Writings 1909–53* (transl. E. H. Schneewind), eds. M. Reichenbach and R. S. Cohen. Reidel, pp. 81–225.

Reszka, R. (1998). Liposomes as drug carrier for diagnostics, cytostatics and genetic material. In *Future Strategies for Drug Delivery with Particulate Systems*, eds. J. E. Diederichs and R. H. Müller. Medpharm GmbH Scientific Publishers.

Ribo, J. M., Crusats, J., Sagues, F., Claret, J., and Rubires, R. (2001). Chiral sign induction during the formation of mesophases in stirred solutions. *Science*, **292**, 2063–6.

Rikken, G. L. and Raupach, E. (2000). Enantioselective magnetochiral photochemistry. *Nature*, **405**, 895–6.

Rispens, T. and Engberts, J. B. F. N. (2001). Efficient catalysis of a Diels-Alder reaction by metallo-vesicles in aqueous solution. *Org. Lett.*, **3**, 941–943.

Riste, T. and Sherrington, D., ed. (1996). *Physics of Biomaterials: Fluctuations, Selfassembly and Evolution* (Nato Science Series Series E, Applied Sciences). Kluwer.

Rizzotti, M., ed. (1996). *Defining Life*. University of Padua.

Robertson, R. N. (1983). *The Lively Membrane*. Cambridge University Press.

Rode, B. M., Son, H. L. and Suwannachot, Y. (1999). The combination of salt induced peptide formation reaction and clay catalysis: a way to higher peptides under primitive earth conditions. *Orig. Life. Evol. Biosph.*, **29**, 273–86.

Rolle, F. (1863). *Ch. Darwin's Lehre von der Entstehung der Arten, in ihrer Anwendung auf die Schöpfunggeschichte*. J. C. Hermann.

Roseman, A., Lentz, B. R., Sears, B., Gibbes, D., and Thompson, T. E. (1978). Properties of sonicated vesicles of three synthetic phospholipids. *Chem. Phys. Lipids*, **21**, 205–22.

Rotello, V., Hong, J. I., and Rebek, J. (1991). Sigmoidal growth in a self-replicating system. *J. Am. Chem. Soc.*, **113**, 9422–3.

Roux, A., Cappello, G., Carteaud, J., *et al.* (2002). A minimal system allowing tubulation with molecular motors pulling on giant liposomes. *Proc. Natl. Acad. Sci.*, **99**, 5394–9.

Rushdi, A. I. and Simoneit, B. R. (2001). Lipid formation by aqueous Fischer-Tropf-type synthesis over a temperature range 100 to 400 °C. *Orig. Life Evol. Biosph.*, **31**, 103–18.

Sackmann, E. (1978). Dynamic molecular-organization in vesicles and membranes. *Ber. Bunsen-Gesell. Phys. Chem.*, **82**, 891–909.

  (1995). In *Structure and Dynamics of Membranes*, eds. R. Lipowsky and E. Sackmann. Elsevier Science, vol 1, pp. 213–304.

Sada, E., Katoh, S., Terashima, M., and Tsukiyama, K.-I. (1988). Entrapment of an ion-dependent enzyme into reverse-phase evaporation vesicles. *Biotechnol. Bioeng.*, **32**, 826–30.

Sada, E., Katoh, S., Terashima, M., Shiraga, H., and Miura, Y. (1990). Stability and reaction characteristics of reverse-phase evaporation vesicles (revs) as enzyme containers. *Biotechnol. Bioeng.*, **36**, 665–71.

Saetia, S., Liedl, K. R., Eder, A. H., and Rode, B. M. (1993). Evaporation cycle experiments: a simulation of salt-induced peptide synthesis under possible prebiotic conditions. *Orig. Life Evol. Biosph.*, **3**, 167–76.

Sagan, C. (1994). The search for extraterrestrial life. *Sci. Amer.*, **271** (4), 71–77.

  (1985). *Cosmos*. Ballantine Publishing.

Sanchez, R. A., Ferris, J. P., and Orgel, L. E. (1966). Conditions for purine synthesis: did prebiotic synthesis occur at low temperature? *Science*, **153**, 72–3.

  (1968). Studies in prebiotic synthesis. IV, The conversion of 4-aminoimidazole-5-carbonitrile derivatives to purines. *J. Mol. Biol.*, **38**, 121–8.

Sankararaman, S., Menon, G. I., and Kumar, P. B. S. (2004). Self-organized pattern formation in motor-microtubule mixtures. *Phys. Rev.*, **70**, 31904–18.

Sato, I., Kadowaki, K., Ohgo, Y., and Soai, K. (2004). Highly enantio selective asymmetric autocatalysis induced by chiral ionic crystals of sodium chlorate and sodium bromate. *J. Mol. Catal. A*, **216**, 209–14.

Scartazzini, R. and Luisi, P. L. (1988). Organogels from lecithins. *J. Phys. Chem.*, **92**, 829–33.

Schaerer, A. A. (2002). Conceptual conditions for conceiving life – a solution for grasping its principle, not mere appearances. In *Fundamentals of Life*, eds. G. Palyi, C. Zucchi, and L. Caglioti. Elsevier, pp. 589–624.

Schmidli, P. K., Schurtenberger, P., and Luisi, P. L. (1991). Liposome-mediated enzymatic synthesis of phosphatidylcholine as an approach to self-replicating liposomes. *J. Am. Chem. Soc.*, **113**, 8127–30.

Schopf, J. W. (1992). In *The Proterozoic Atmosphere*, eds. J. W. Schopf and C. Klein. Cambridge University Press.

Schopf, J. W. (1993). Microfossils of the early archean apex chert: new evidence of the antiquity of life. *Science*, **260**, 640–6.

Schopf, J. W. (1998). In *The Molecular Origin of Life*, ed. A. Brack. Cambridge University Press.

(2002). *Life's Origin*. University of California Press.

Schröder, J. (1998). Emergence: non-deducibility or downward causation? *Phil. Q.*, **48**, 434–52.

Schurtenberger, P., Magid, L. J., King, S. M. and Lindner, P. (1991). Cylindrical structure and flexibility of polymerlike lecithin reverse micelles. *J. Phys. Chem.*, **95**, 4173–6.

Schurtenberger, P., Scartazzini, R., Magid, L. J., Leser, M. E., and Luisi, P. L. (1990). Structural and dynamic properties of polymer-like reverse micelles. *J. Phys. Chem.*, **94**, 3695–701.

Schuster, P. and Swetina, J. (1988). Stationary mutant distributions and evolutionary optimization. *Bull. Math. Biol.*, **50**, 636–60.

Schwabe, C. (2001). *The Genomic Potential Hypothesis, a Chemist's View on the Origin and Evolution of Life*. Landes Bioscience.

(2002). Genomic potential hypothesis of evolution: a concept of biogenesis in habitable spaces of the universe. *Anat. Rec.*, **268**, 171–9.

(2004). Chemistry and biodiversity: Darwinism, evolution, and speciation, *Chem. Biodiv.*, **1**, 1588–90.

Schwabe, C. and Warr, G. W. (1984). A polyphyletic view of evolution. The genetic potential hypothesis. *Persp Biol. Med.*, **27**, 465–85.

Schwartz, A. W. (1998). Origin of the RNA world. In *The Molecular Origins of Life*, ed. A. Brack. Cambridge University Press.

Schwendiger, M. G. and Rode, B. M. (1992). Investigations on the mechanism of the salt-induced peptide formation. *Orig. Life Evol. Biosph.*, **6**, 349–59.

Scott, E. (2004). *Evolution vs. Creationism: An Introduction*. Greenwood Press. See scott@natcenscied.org; www.ncseweb.org.

Seddon, J. M. (1990). Structure of the inverted hexagonal (HII) phase, and non-lamellar phase transitions of lipids, *Biochim. Biophys. Acta*, **1031**, 1–69.

Seddon, J. M., Hogan, J. J., Warrender, N. A., and Pebay-Peyroula, E. (1990). *Prog. Coll. Polym. Sci.*, **81**, 189–97.

Segre, D. and Lancet, D. (2000). Composing life. *EMBO Rep.*, **1** (3), 217–22.

Segre, D., Ben-Eli, D., and Lancet, D. (2000). Compositional genomes: prebiotic information transfer in mutually catalytic non-covalent assemblies. *Proc. Natl. Acad. Sci. USA*, **97** (8), 4112–17.

Segre, D., Ben-Eli, D., Deamer, D., and Lancet, D. (2001). The lipid world. *Orig. Life Evol. Biosph.*, **31**, 119–45.

Severin, K., Lee, D. H., Kennan, A. J., and Ghadiri, M. R. (1997). A synthetic peptide ligase. *Nature*, **16** (389), 706–9.

Shapiro, R. (1986). *Origins: a Skeptic's Guide to the Creation of Life on Earth*. Summit Books.

(1988). Prebiotic ribose synthesis: a critical analysis. *Orig. Life Evol. Biosph.*, **18**, 71–85.

(1995). The prebiotic role of adenine: a critical analysis. *Orig. Life Evol. Biosph.*, **25**, 83–98.

Shen, C., Lazcano, A., and Oró, J. (1990a). The enhancement activities of histidyl-histidine in some prebiotic reactions. *J. Mol. Evol.*, **31**, 445–52.

Shen, C., Mills, T., and Oró, J. (1990b). Prebiotic synthesis of histidyl-histidine. *J. Mol. Evol.*, **31**, 175–9.

Shenhav, B. and Lancet, D. (2004). Prospects of a computational origin-of-life endeavor. *Orig. Life Evol. Biosph.*, **34**, 181–94.

Shermer, M. (2003). Is the universe fine-tuned for life? *Sci. Amer.*, Jan, 23.

Shimkets, L. J. (1998). Structure and sizes of genomes of the archaea and bacteria. In *Bacterial Genomes: Physical Structure and Analysis*, eds. F. J. De Bruijn, J. R. Lupskin, and G. M. Weinstock. Kluwer, pp. 5–11.

Shiner, E. K., Rumbaugh, K. P., and Williams, S. C. (2005). Interkingdom signaling: deciphering the language of acyl homoserine lactones. *FEMS Microbiol. Rev.*, **29**, 935–47.

Shostak, S. (2003). *Panspermia: Spreading Life Through the Universe*. The SETI Institute.

Sievers, D., Achilles, T., Burmeister, J., *et al.* (1994). In *Self-Production of Supramolecular Structures*, eds. G. Fleischacker, S. Colonna and P. L. Luisi. Kluwer Publishers.

Sievers, D. and von Kiedrowski, G. (1994). Self-replication of complemenary nucleotide-based oligomers. *Nature*, **369**, 221–4.

Silin, V. I., Wieder, H., Woodward, J. T. *et al.* (2002). The role of surface free energy on the formation of hybrid bilayer membranes. *J. Am. Chem. Soc.*, **124**, 14676–83.

Simon, H. A. (1969). *The Sciences of the Artificial*. MIT Press.

Simpson, G. G. (1973). Added comments on "The non-prevalence of humanoids". In *Communication with Extraterrestrial Intelligence*, ed. C. Sagan. MIT Press, pp. 362–4.

Smith, H. O., Hutchison, C. A. III, Pfannkoch, C., and Venter, J. C. (2003). Generating a synthetic genome by whole genome assembly: phiX174 bacteriophage from synthetic oligonucleotides. *Proc. Natl. Acad. Sci. USA*, **100**, 15440–5.

Smith, R. S. and Iglewski, B. H. (2003). *P. aeruginosa* quorum sensing systems and virulence. *Curr. Opin. Microbiol.*, **6**, 56–60.

Solomon, B. and Miller, I. R. (1976). Interaction of glucose oxidase with phospholipid vesicles. *Biochim. Biophys. Acta*, **455**, 332–42.

Sperry, R. W. (1986). Discussions: Macro- versus Microdeterminism. *Phil. Sci.*, **53**, 265–70.

Spirin, A. ((1986)). *Ribosome Structure and Protein Synthesis*. Benjamin Cummings Publishing.

Stano, P., Bufali, S., Pisano, C., *et al.* (2004). Novel campotothecin analogue (Gimatecan)-containing liposomes prepared by the ethanol injection method. *J. Lipos. Res.*, **14**, 87–109.

Stano, P., Wehrli, E., and Luisi, P. L. (in press). Insights on the self-reproduction of oleate vesicles, *J. Phys. Condensed Matter*.

Stetter, K. O. (1998). Hyperthermophiles and their possible role as ancestors of modern life. In *The Molecular Origin of Life*, ed. A. Brack. Cambridge University Press.

Stocks, P. G. and Schwarz, A. W. (1982). Basic nitrogen-heterocyclic compounds in the Murchison meteorite. *Geochim. Cosmochim. Acta*, **46**, 309–15.

Strogatz, S. H. (1994). *Non Linear Dynamics and Chaos, With Applications to Physics, Biology, Chemistry, and Engineering*. Perseus Book Group.

Stryer, L. (1975). *Biochemistry*, Freeman and Co.

Suttle, D. P. and Ravel, J. M. (1974). The effects of initiation factor 3 on the formation of 30S initiation complexes with synthetic and natural messengers. *Biochem. Biophys. Res. Commun.*, **57**, 386–93.

Suwannachot, Y. and Rode, B. M. (1999). Mutual amino acid catalysis in salt-induced peptide formation supports this mechanism's role in prebiotic peptide evolution. *Origin Life Evol. Biosph.*, **5**, 463–71.

Swairjo, M. A., Seaton, B. A., and Roberts, M. F. (1994). *Biochem. Biophys. Acta*, **1191**, 354–61.

Szathmáry, E. (2002). Units of evolution and units of life. In *Fundamentals of Life*, eds. G. Palyi, L. Zucchi and L. Caglioti. Elsevier SAS, pp. 181–95.

Szostak, J. W., Bartel, D. P., and Luisi P. L. (2001). Synthesizing life. *Nature*, **409**, 387–90.

Taillades, J., Cottet, H., Garrel, L., *et al.* (1999). N-Carbamoyl amino acid solid–gas nitrosation by NO/NOx: a new route to oligopeptides via α-amino acid N-carboxyanhydride. Prebiotic implications. *J. Mol. Evol.*, **48**, 638–45.

Takahashi, Y. and Mihara, H. (2004). Construction of a chemically and conformationally self-replicating system of amyloid-like fibrils. *Bioorg. Med. Chem.*, **12**, 693–9.

Takakura, K., Toyota, T., and Sugawara, T. (2003). A novel system of self-reproducing giant vesicles. *J. Am. Chem. Soc.*, **125**, 8134–8140. See also: Takakura, K., Toyota, T., Yamada, K., *et al.* (2002). Morphological change of giant vesicles triggered by dehydrocondensation reaction. *ChemLett.*, **31**, 404–5.

Tanford, C. (1978). The hydrophobic effect and the organization of living matter. *Science*, **200**, 1012–18.

Teramoto, N., Imanishi, Y., and Yoshihiro, I. (2000). In vitro selection of a ligase ribozyme carrying alkylamino groups in the side chains. *Bioconjugate Chem.*, **11**, 744–8.

Theng, B. K. G. (1974). *The Chemistry of Clay-Organic Reactions*. Adam Hilger.

Thomas, C. F. and Luisi, P. L. (2005). RNA selectively interacts with vesicles depending on their size. *J. Phys. Chem. B.*, **109**, 14544–50.

Thomas, C. F. and Luisi, P. L. (2004). Novel properties of DDAB: matrix effect and interaction with oleate. *J. Phys. Chem. B*, **108**, 11285–90.

Thomas, P. J., Chiba, C. F., and McKay, C. P., eds. (1997a). *Comets and the Origins and Evolution of Life*. Springer Verlag.

Thomas, R. M., Wendt, H., Zampieri, A., and Bosshard, H. R. (1995). Alpha-helical coiled coils: simple models for self-associating peptide and protein systems. *Prog. Coll. Polym. Sci.*, **99**, 24–30.

Thomas, R. M., Zampieri, A., Jumel, K., and Harding, S. E. (1997b). A trimeric, alpha-helical, coiled coil peptide: association stoichiometry and interaction strength by analytical ultracentrifugation. *Eur. Biophys. J.*, **25**, 405–10.

Thompson, E. and Varela, F. J. (2001a). Radical embodiment: neural dynamics and consciousness. *Trends Cog. Sci.*, **5**, 418–25.

Tjivikua, T., Ballister, P., and Rebek, J. (1990). A self-replicating system. *J. Am. Chem. Soc.*, **112**, 1249–50.

Tranter, G. E. (1985a). The parity-violating energy difference between enantiomeric reactions. *Chem. Phys. Lett.*, **115**, 286.

(1985b). The effects of parity violation on molecular structure. *Chem. Phys. Lett.*, **121**, 339.

Tranter, G. E., MacDermott, A. J., Overill, R. E., and Speers, P. (1992). Computational studies of the electroweak origin of biomolecular handedness in natural sugars. *Proc. Royal Soc., London A*, **436**, 603–15.

Trinks, H., Schroeder, W., and Biebricher, C. K. (2003). *Eis und die Entstehung des Lebens (Ice and the Origin of Life)*. Shaker Verlag.

Tsukahara, H., Imai, E. I., Honda, H., Hatori, K., and Matsuno, K. (2002). Prebiotic oligomerization on or inside lipid vesicles in hydrothermal environments. *Orig. Life Evol. Biosph.*, **32**, 13–21.

Tsumoto, K., Nomura, S. M., Nakatani, Y., and Yoshikawa, K. (2002). Giant liposome as a biochemical reactor: transcription of DNA and transportation by laser tweezers. *Langmuir*, **17**, 7225–8.

Turing, A. (1952). The chemical basis of Morphogenesis. *Phil. Trans. Royal. Soc. London B*, **237**, 37.

Ulbricht, W. and Hoffmann, H. (1993). Physikalische Chemie der Tenside. In *Die Tenside*, ed. K. Kosswig, and H. Stache. Carl Hanser Verlag, pp. 1–114.

Ulman, A. (1996). Formation and structure of self-assembled monolayers. *Chem. Rev.*, **96**, 1533–54.

Uster, P. S. and Deamer, D. W. (1981). Fusion competence of phosphatidylserine-containing liposomes quantitatively measured by a fluorescence resonance energy transfer assay. *Arch. Biochem. Biophys.*, **209** (2), 385–95.

Vaida, M., Popovitz-Biro, R., Leiserowitz, L., and Lahav, M. (1991). Probing reaction pathways via asymmetric transformations in chiral and centrosymmetric crystals. In *Photochemistry in Organized and Condensed Media*, ed. V. Ramamurthy. VCH, pp. 248–302.

Valenzuela, C. Y. (2002). Does biotic life exist. In *Fundamentals of Life*, eds. G. Palyi, C. Zucchi, and L. Cagiliati. Elsevier, pp. 331–4.

Varela, F. J. (1979). *Principles of Biological Autonomy*. North Holland/Elsevier.
  (1989a). Reflections on the circulation of concepts between a biology of cognition and systemic family therapy. *Family Process*, **28**, 15–24.
  (1989b). *Autonomie et Connaissance*. Seuil, p. 167.

Varela, F. J. (1999). *Ethical Know-How: Action, Wisdom, and Cognition*. Stanford University Press.
  (2000). *El Fenomeno de la Vita*. Dolmen Ensayo.

Varela, F. J., Maturana, H. R., and Uribe, R. B. (1974). Autopoiesis: the organization of living system, its characterization and a model. *Biosystems*, **5**, 187–96.

Varela, F. J., Thompson, E., and Rosch, E. (1991). *The Embodied Mind*. MIT Press.

Wächtershäuser, G. (1988). Before enzymes and templates: theory of surface metabolism. *Microbiol. Rev.* **52**, 452–84.
  (1990a). Evolution of the first metabolic cycles. *Proc. Natl. Acad. Sci. USA*, **87**, 200–4.
  (1990b). The case for the chemoautotrophic origin of life in the iron–sulfur world. *Origin Life Evol. Biosph.*, **20**, 173–6.
  (1992). Groundworks for an evolutionary biochemistry: the iron–sulfur world. *Prog. Biophys. Mol. Biol.*, **58**, 85–201.
  (1997). The origin of life and its methodological challenge. *J. Theor. Biol.*, **187**, 483–94.
  (2000). Life as we don't know it. *Science*, **289**, 1307–8.

Waks, M. (1986) Proteins and peptides in water-restricted environments. *Proteins*, **1**, 4–15.

Walde, P. (2000). Enzymatic reactions in giant vesicles. In: *Giant Vesicles, Perspectives in Supramolecular Chemistry*, eds. P. L. Luisi and P. Walde. John Wiley & Sons Ltd., pp. 297–311.

Walde, P. and Ichikawa, S. (2001). Enzymes inside lipid vesicles: preparation, reactivity and applications. *Biomol. Eng.*, **18**, 143–77.

Walde P. and Luisi P. L. (1989). A continuous assay for lipases in reverse micelles based on fourier transform infrared spectroscopy. *Biochemistry*, **28**, 3353–7.

Walde, P. and Mazzetta, B. (1998) Bilayer permeability-based substrate selectivity of an enzyme in liposomes. *Biotechnol. Bioeng.*, **57**, 216–19.

Walde, P., Goto, A., Monnard, P.-A., Wessicken, M., and Luisi, P. L. (1994a). Oparin's reactions revisited: enzymatic synthesis of poly(adenylic acid) in micelles and self-reproducing vesicles. *J. Am. Chem. Soc.*, **116**, 7541–7.

Walde, P., Wick, R., Fresta, M., Mangone, A., and Luisi, P. L. (1994b). Autopoietic self-reproduction of fatty acid vesicles. *J. Am. Chem. Soc.*, **116**, 11649–54.

Weaver, W. (1948). Science and complexity. *Amer. Sci.*, **36**, 536–44.

Weber, A. (2002). The 'surplus of meaning'. Biosemiotic aspects in Francisco J. Varela's philosophy of cognition. *Cybernetics Human Knowing*, **9**, 11–29.

Weiner, A. M. and Maizels, N. (1987). tRNA-Like structures tag the 3′ ends of genomic RNA molecules for replication: implications for the origin of protein synthesis. *Proc. Natl. Acad. Sci. USA*, **84**, 7383–7.

Weissbuch, I., Popovitz-Biro, R., Leizerowitz, L., and Lahav, M. (1994). In *The Lock-and the Key Principle The State of the Art-100 Years On*, 1, ed. J.-P. Behr. John Wiley & Sons Ltd., p. 173 and references cited therein.

Weissbuch, I., Frolow, F., Addadi, L., Lahav, M., and Leiserowitz, L. (1990). Oriented crystallization as a tool for detecting ordered aggregates of water-soluble hydrophobic alfa-amino acids at the air-solution interface. *J. Am. Chem. Soc.*, **112**, 7718–24.

Weissbuch, I., Addadi, L., Berkovitch-Yellin, Z., *et al.* (1984). Spontaneous generation and amplification of optical activity in amino acids by enantioselective occlusion into centrosymmetric crystals of glycine. *Nature*, **310**, 161–4.

Weissbuch, I., Zepik, H., Bolbach, G., *et al.* (2003). Homochiral oligopeptides by chiral amplification within two-dimensional crystalline self-assemblies at the air–water interface; relevance to biomolecular handedness. *Chemistry*, **9** (8), 1782–94.

Wendt, H., Durr, E., Thomas, R. M., Przybylski, M., and Bosshard, H. R. (1995). Characterization of leucine zipper complexes by electrospray ionization mass spectrometry. *Protein Sci.*, **4**, 1563–70.

Wenneström, H. and Lindmann, B. (1979). *Phys. Rev.*, **52**, 1–86.

Westhof, E. and Hardy, N., eds. (2004). *Folding and Self-Assembly of Biological Macromolecules*. World Scientific Publishing Company Inc.

Whitesides, G. M. and Boncheva, M. (2002). Beyond molecules: self-assembly of mesoscopic and macroscopic components. *Proc. Natl. Acad. Sci. USA*, **99**, 4769–74.

Whitesides, G. M., Mathias, J. P., and Seto, C. T. (1991). Molecular self-assembly and nanochemistry – a chemical strategy for the synthesis of nanostructures. *Science*, **254**, 1312–19.

Wick, R., Walde, P., and Luisi, P. L. (1995). Autocatalytic self-reproduction of giant vesicles. *J. Am. Chem. Soc.*, **117**, 1435–6.

Willimann, H. and Luisi, P. L. (1991). Lecithin organogels as matrix for the transdermal transport of drugs. *Biochem. Biophys. Res. Commun.*, **177**, 897–900.

Wilschut, J., Duzgunes, N., Fraley, R., and Papahadjopoulos, D. (1980). Studies on the mechanism of membrane-fusion – kinetics of calcium-ion induced fusion of phosphatidylserine vesicles followed by a new assay for mixing of aqueous vesicle contents. *Biochemistry*, **19**, 6011–21.

Wilson, T. L. (2001). The search for extraterrestrial intelligence. *Nature*, **409**, 1110–14.

Wimsatt, W. C. (1972). Complexity and organization. In *Boston Studies in the Philosophy of Science, Proceedings of the Philosphy of Science Association*, eds. K. F. Schaffner and R. S. Cohen. Reidel, pp. 67–86.

   (1976a). Reductionism, levels of organization, and the mind-body problem. In *Consciousness and the Brain*, eds. G. Globus, G. Maxwell, and I. Savodinik. Plenum Press, pp. 205–66.

(1976b). Reductive explanation, a functional account. In *Proceedings of the Meetings of the Philosophy of Science Association 1974*, eds. C. A. Hooker, G. Pearse, A. C. Michealos, and J. W. van Evra Reidel, pp. 671–710.

Winfree, A. T. (1984). The prehistory of the Belousov-Zhabotinsky oscillator. *J. Chem. Educ.*, **61**, 661–3.

Woese, C. R. (1979) A proposal concerning the origin of life on the planet Earth. *J. Mol. Evol.* **13**, 95–101.

(1983). The primary lines of descent and the universal ancestor. In *Evolution from Molecules to Man*, ed. D. S. Bendall. Cambridge Universiy Press, pp. 209–33.

Wong, J. T.-F., Xue, H. (2002). Self-perfecting evolution of heteropolymer building blocks and sequences as the basis of life. In *Fundamentals of Life*, eds. G. Pályi, L. Zucchi, and L. Caglioti. Elsevier SAS, pp. 473–94.

Wood, W. B. (1973). Genetic control of bacteriophage T4 morphogenesis. In *Genetic Mechanisms of Development*, ed. F. J. Ruddle. Academic Press, pp. 29–46.

Woodle, M. C. and Lasic, D. D. (1992). *Biochim. Biophys. Acta*, **1113**, 171–99.

Yao, S., Ghosh, I., and Chmielewski, J. (1998). Selective amplification by auto- and cross-catalysis in a replicating peptide system. *Nature*, **396**, 447–50.

Yao, S., Ghosh, I., Zutshi, R., and Chmielewski, J. (1997). A pH-modulated self-replicating peptide. *J. Am. Chem. Soc.*, **119**, 10559–60.

Yaroslavov, A. A., Udalyk, O. Y., Kabanov, V. A., and Menger, F. M. (1997). Manipulation of electric charge on vesicles by means of ionic surfactants: effects of charge on vesicle mobility, integrity, and lipid dynamics. *Chem. Eur. J.*, **3**, 690–5.

Yoshimoto, M., Walde, P., Umakoshi, H., and Kuboi, R. (1999). Conformationally changed cytochrome c-mediated fusion of enzyme- and substrate-containing liposomes. *Biotechnol. Prog.*, **15**, 689–96.

Yu, W., Sato, K., Wakabayashi, M., *et al.* (2001). Synthesis of functional protein in liposome. *J. Biosc. Bioeng.*, **92**, 590–3.

Yuen, G. U. and Knevolden, K. A. (1973). Monocarboxylic acids in Murray and Murchison carbonaceous meteorites. *Nature*, **246**, 301–2.

Yuen, G. U., Lawless J. G., and Edelson, E. H. (1981). Quantification of monocarboxylic acids from a spark discharge synthesis. *J. Mol. Evol.*, **17**, 43–7.

Zamarev, K. I., Romannikov, V. N., Salganik, R. I., Wlassoff, W. A., and Khramtsov, V. V. (1997). Modelling of the prebiotic synthesis of oligopeptides: silicate catalysts help to overcome the critical stage. *Orig. Life Evol. Biosph.*, **27**, 325–37.

Zampieri, G. G., Jäckle, H., and Luisi, P. L. (1986). Determination of the structural parameters of reverse micelles after uptake of proteins. *J. Phys. Chem.*, **90**, 1849.

Zeleny, M. (1977). Self-organization of living systems formal model of autopoiesis. *Int. J. Gen. Syst.*, **4**, 13–28.

Zelinski, W. S. and Orgel, L. E. (1987). Autocatalytic synthesis of a tetranucleotide analogue. *Nature*, **327**, 346–7.

Zeng, F. W. and Zimmermann, S. C. (1997). Dendrimers in supramolecular chemistry: from molecular recognition to self-assembly. *Chem. Rev.*, **97**, 1681–712.

Zeng, X., Ungar, G., Liu, Y., *et al.* (2004). Supramolecular dendritic liquid quasicrystals. *Nature*, **428**, 157–60.

Zepik, H. H., Bloechliger, E., and Luisi, P. L. (2001). A chemical model of homeostasis. *Angew. Chem. Int. Ed. Engl.*, **40**, 199–202.

Zhang, B. and Cech, T. R. (1998). Peptidyl-transferase ribozymes: trans reactions, structural characterization and ribosomal RNA-like features. *Chem. Biol.*, **5**, 539–53.

Zhao, M. and Bada, J. L. (1989). Extraterrestrial amino acids in cretaceous/tertiary boundary sediments at Stevns Klint, Denmark. *Nature*, **339**, 463–5.

Zhu, J., Zhang, L., and Reszka, R. (1996a). Liposome-mediated delivery of genes and oligonucleotides for the treatment of brain tumors. In *Targeting of Drugs: Strategies for Oligonucleide and Gene Delivery in Therapy*, eds. G. Gregoriadis and B. McCormack. Plenum Press, pp. 169–87.

Zhu, J., Zhang, L., Hanisch, U. K., Felgner, P. L., and Reszka, R. (1996b). In vivo gene therapy of experimental brain tumors by continuous administration of DNA-liposome complexes. *Gene Therapy*, **3**, 472–6.

Ziegler, M., Davis, A. V., Johnson, D. W., and Raymond, K. N. (2003). Supramolecular chirality: a reporter of structural memory. *Angew. Chem. Int. Ed. Engl.*, **42**, 665–8.

Zimmer, C. (2003). Tinker, tailor: can Venter stitch together a genome from scratch? *Science*, **299**, 1006–8.

Index

301

Lightning Source UK Ltd.
Milton Keynes UK
UKOW020904140513

210617UK00007B/278/P